Igor V. Novozhilov

Fractional Analysis

Methods of Motion Decomposition

1997
Birkhäuser
Boston • Basel • Berlin

Igor V. Novozhilov
Department of Mechanics
and Mathematics
Moscow State University
Moscow 119899
Russia

Library of Congress Cataloging-in-Publication Data

Novozhilov, I. V. (Igor ' Vasil ' evich)
[Fraktsionnyĭ analiz. English]
Fractional analysis : methods of motion decomposition /
I. V. Novozhilov.
p. cm.
Includes bibliographical references and index.
ISBN-13:978-1-4612-8667-7 e-ISBN-13:978-1-4612-4130-0
DOI:10.1007/978-1-4612-4130-0

1. Differential equations--Numerical solutions. 2. Differentiable dynamical systems. 3. Approximation theory. I. Title.
QA372.N6813 1997
531 '.11 '01515352--DC21 96-49997
CIP

Printed on acid-free paper

Birkhäuser
Softcover reprint of the hardcover 1st edition 1997

Typeset by the Author in LaTeX.
Printed and bound by Quinn-Woodbine, Woodbine, NJ.

9 8 7 6 5 4 3 2 1

Contents

Preface

This book considers methods of approximate analysis of mechanical, electromechanical, and other systems described by ordinary differential equations.

Modern mathematical modeling of sophisticated mechanical systems consists of several stages: first, construction of a mechanical model, and then writing appropriate equations and their analytical or numerical examination. Usually, this procedure is repeated several times. Even if an initial model correctly reflects the main properties of a phenomenon, it describes, as a rule, many unnecessary details that make equations of motion too complicated. As experience and experimental data are accumulated, the researcher considers simpler models and simplifies the equations. Thus some terms are discarded, the order of the equations is lowered, and so on. This process requires time, experimentation, and the researcher's intuition. A good example of such a semi-experimental way of simplifying is a gyroscopic precession equation. Formal mathematical proofs of its admissibility appeared some several decades after its successful introduction in engineering calculations.

Applied mathematics now has at its disposal many methods of approximate analysis of differential equations. Application of these methods could shorten and formalize the procedure of simplifying the equations and, thus, of constructing approximate motion models.

Wide application of the methods into practice is hindered by the following.

1. Descriptions of various approximate methods are scattered over the mathematical literature. The researcher, as a rule, does not know what method is most suitable for a specific case.

2. The preceding methods are essentially variants of the method of small parameter. Equations of systems for the most part do not contain small parameters. The problem arises of reducing the initial equations to a form containing small parameters so that mathematical formalism can be applied to them. The problem is solved using similitude theory.

The two stages, introduction of small parameter and application of mathematical tools, together form a procedure which has been termed fractional analysis. Fractional analysis singles out the main motion components and small "additives" to them, and separates slow and fast components, in short, separates the motion on big and small fractions.

Such "fractioning" facilitates the numerical analysis of dynamic systems. It is known that one of the difficulties of numerical analysis is separation of eigenfrequencies of a system when one has to take into account high-frequency components over large time intervals with a small integration step. Methods of fractional analysis make it possible to obtain approximate equations that describe fast and slow components of motion separately. Each of these equations can be solved numerically on its own time scale.

The book presents basics of fractional analysis. It can be of interest to students, postgraduates, and researchers who deal with analytical and numerical analysis of equations of motion of dynamic systems. The book surveys the variety of existing methods of small parameters and shows how and where to use them. The author tries to present the material in an easy-to-understand way, omitting proofs of mathematical theorems and simplifying mathematical formulations.

The content of the book was certainly affected by the author's taste and experience. He has been dealing mainly with "rough" systems that are presented ambiguously by mechanical models, are difficult to be described formally, and possess high damping. Approximate analysis of such objects is preferably done by the method of successive approximations, boundary-layer methods, and, to a lesser extent, by the method of averaging. The author does not deal with pure Hamiltonian mechanics with its celestial-mechanical and trajectory-ballistic applications. Hence the bulk of resonance situations of modern methods of averaging are outside the scope of the book.

The book is to fill the gap between mathematical formalism and applications. Hence emphasis is on examples, mainly from the mechanics of controlled motion. In them, the whole cycle of fractional analysis is performed, from writing initial equations and introducing small parameters to constructing models and estimating their accuracy. The author's own results on construction and analysis of approximate mathematical models in the mechanics of gyroscopes, transport, robots, and the like, are presented.

CHAPTER I

Dimensional analysis and small parameters

1 Dimensional analysis

1.1 The main concepts of dimensional analysis

Scientific disciplines deal with quantities that are by nature qualitatively different and that are measured with the help of numbers.

In analytical description qualitative concepts are represented by different names or labels, that is, dimensions. For example, there exist quantities with dimensions such as "length," "mass," "energy," and so on. To make a qualitative idea of a physical characteristic quantitative, a unit and procedure of measurement must be defined. Thus, the meter (or its fraction) is a unit of length. Therefore, the length of a beam is measured by the number of times a given scale unit (that is, the meter and its fractions) can be laid off along the beam.

Let a length of a beam and a unit of length be denoted by L and L_*, respectively. Then $L = L_*\ell$, where L and L_* are quantities with dimension "length," and ℓ is a numerical measure (that is, relative number) that determines the length of a beam for given L_*. An area S, velocity V, time T, and any other dimensional quantity can be determined similarly. Therefore,

$$L = L_*\ell, \quad S = S_*s, \quad V = V_*v, \quad T = T_*t, \quad \dots , \tag{1.1}$$

where $L_*, S_*, V_*, T_*, \dots$ are units of measurement, and $\ell, s, v, t, \dots$ are numerical measures of corresponding quantities.

Here and in the remainder of this treatment quantities with dimensions are denoted by capital letters and their numerical measures are denoted by small letters.

In any specific problem a set of quantities, variables, and constants is used. The problem of choosing these quantities is related to the problem of choosing a mathematical model proper for the phenomenon under consideration. If a model is already determined, that is, all the equations and initial and boundary conditions are already written, then this set consists of all the notations involved in the model representation.

The quantities mentioned previously must satisfy a number of base relations, which were established for problems with the same set of variables and constants. These relations are usually written as definitions of one quantity by means of other ones.

In mechanics the following base definitions are usually accepted: definition of an area of a square (S) via its side length (L); definition of speed velocity (V) as the rate of change of length with respect to time; definition of acceleration W as the rate of change of velocity with respect to time; Newton's Second Law, which is often considered as a definition of a force F by means of acceleration and mass M; and so on.

The preceding relations can be written as

$$S = L^2, \quad V = \frac{dL}{dT}, \quad W = \frac{dV}{dT}, \quad F = MW, \ \ldots \ . \tag{1.2}$$

From the point of view of dimensional analysis, the quantities $L, S, T, \ldots$ and relations (1.2) comply with two almost trivial postulates [3, 4].

1. Quantities $L, S, V, \ldots$ themselves do not depend on the choice of units of measurement and the measurement procedure.

2. Only quantities with the same dimensions can be used in such relations as equality or inequality.

After conversion in (1.2) from dimensional quantities to relative quantities, with numerical measures defined by (1.1), one can obtain

$$S_*s = L_*^2\ell^2, \quad V_*v = \frac{L_*}{T_*}\frac{d\ell}{dt}, \quad W_*w = \frac{V_*}{T_*}\frac{dv}{dt}, \quad F_*f = M_*W_*mw, \dots ,$$

or

$$s = \Delta_1\ell^2, \quad v = \Delta_2\frac{d\ell}{dt}, \quad w = \Delta_3\frac{dv}{dt}, \quad f = \Delta_4 mw, \quad \dots , \tag{1.3}$$

where

$$\Delta_1 = \frac{L_*^2}{S_*}, \quad \Delta_2 = \frac{L_*}{T_*V_*}, \quad \Delta_3 = \frac{V_*}{T_*W_*}, \quad \Delta_4 = \frac{M_*W_*}{F_*}, \quad \dots . \tag{1.4}$$

The quantities $s, \ell, \dots$ in (1.3) are relative real numbers. So, the second postulate is satisfied only if $\Delta_1, \Delta_2, \dots$ are also relative quantities. Usually these quantities are chosen to be equal to one: $\Delta_1 = \Delta_2, = \dots = 1$. Then from (1.4) it follows that

$$\frac{L_*^2}{S_*} = 1, \qquad \frac{L_*}{T_*V_*} = 1, \qquad \dots . \tag{1.5}$$

Therefore

$$S_* = L_*^2, \quad V_* = \frac{L_*}{T_*}, \quad W_* = \frac{V_*}{T_*}, \quad F_* = M_*W_*, \quad \dots . \tag{1.6}$$

It is readily seen that the generally accepted definitions are obtained from (1.6). For example, a unit of area is an area of a square with its sides equal to a unit of length L_*, and so on.

Let us suppose that for a problem description there were introduced n units of measurement that satisfy k conditions of the form (1.5). Then $n-k$ units can be chosen to be independent, provided that other k variables are defined uniquely by them.

In mechanics the triples of units (V_*, T_*, M_*), (W_*, L_*, F_*), and so on, can be chosen as independent units of measurement. In the most cases the so-called base system of units L_*, T_*, M_* is chosen as the key triple. The values of these independent units can be arbitrary chosen. For example, in the triple (L_*, T_*, M_*), the meter or the foot can be chosen to be the length unit L_*; the second, the hour, or the year can be chosen to be the time unit T_*.

An expression that determines a dependent measure through independent units is called a dimensional formula. From (1.6) it follows that

$$S_* = L_*^2, \quad V_* = L_*T_*^{-1}, \quad W_* = L_*T_*^{-2}, \quad F_* = M_*L_*T_*^{-2}, \dots . \tag{1.7}$$

It is easily shown that an arbitrary unit dependent with respect to dimensions is a product of powers of independent units. In fact, let a physical quantity Z be expressed by the variables $L, T, \dots$, which are the terms of (1.2), in the form:

$$Z = \Phi(L, T, S, V, \dots). \tag{1.8}$$

Taking (1.1) into account leads to

$$Z_* z = \Phi(L_* \ell, T_* t, S_* s, V_* v, \dots), \tag{1.9}$$

where Z_* is a unit of measure for Z.

Now, by postulate 2, the right-hand side of (1.9) must have a form: $\Phi_* \varphi$, where φ is a relative quantity. This requirement is satisfied if Φ is an homogeneous function of its arguments:

$$Z_* z = L_*^\alpha T_*^\beta S_*^\gamma V_*^\delta \ \dots \ \varphi(\ell, t, s, v, \dots), \tag{1.10}$$

where $\alpha, \beta, \dots$ are exponents of homogeneity.

From (1.10) it follows that $z = \varphi(\ell, t, s, v, \dots)$ and $Z_* = L_*^\alpha T_*^\beta S_*^\gamma V_*^\delta \dots$. Combining this with (1.7) produces

$$Z_* = L_*^{\alpha_1} T_*^{\beta_1} M_*^{\gamma_1}.$$

If all the exponents in a dimensional formula of some quantity are equal to zero, then this quantity is called a dimensionless quantity. An angle being measured in radians is an example of dimensionless quantity. Dimensionless quantity is given by relative number and does not change if the values of the base system units are altered.

Thus, the dimensional structure of a problem (that is, dependent and independent quantities and dimensional formulae) is determined by a set of quantities used in the description of a problem and by the base relations (1.2).

1.2 Transformations in dimensional analysis

Let us analyze how a numerical measure of a base (with respect to dimensions) variable, length, for example, changes if the unit of length L_{1*} is changed for L_{2*}.

From the first postulate of dimensional analysis $L = L_{1*}\ell_1 = L_{2*}\ell_2$. Therefore, the changed numerical measure is $\ell_2 = \lambda \ell_1$, where $\lambda = L_{1*}/L_{2*}$ is a ratio of the size of the initial unit to the size of the changed unit. In the same way,

$$t_2 = \tau t_1, \qquad m_2 = \mu m_1,$$

$$\tau = T_{1*}/T_{2*}, \qquad \mu = M_{1*}/M_{2*}.$$

How are numerical measures of dimension-dependent quantities changed if the values of independent units are altered?

Let, for example, an area, acceleration, or the like be measured with two different sets of base triples: (L_{1*}, T_{1*}, M_{1*}) and (L_{2*}, T_{2*}, M_{2*}). From relations

$$S = S_{1*}s_1 = S_{2*}s_2$$

by substitution of Eqs.(1.7) one can obtain

$$s_2 = \left(\frac{L_{1*}}{L_{2*}}\right)^2 s_1 .$$

In the same way

$$w_2 = \frac{L_{1*}}{L_{2*}} \left(\frac{T_{1*}}{T_{2*}}\right)^{-2} w_1 ,$$

and so on. Consequently,

$$s_2 = \lambda^2 s_1, \qquad w_2 = \lambda\tau^{-2} w_1, \qquad \dots . \tag{1.11}$$

After comparison of (1.11) with (1.7) it is seen that the numerical measure of the dimension-dependent quantity changes according to its dimensional formula.

Left- and right-hand sides of the base relations (1.2) determining a structure of the dimensions of a problem have the same dimensions. This implies that after transition to representation in terms of numerical measures, according to (1.1) and (1.7), the same multiplier appears both in the left-hand and in the right-hand sides of the relations. Therefore, the equations for numerical measures corresponding to Eqs. (1.2) are obtained by renaming the physical quantities $L, T, S, \dots$ with their numerical measures $\ell, t, s, \dots$. Because the dimension of summands in these equations is the same, it follows that if the equations are written in terms of numerical measures, they are invariant with respect to any change of independent-unit values. These statements are valid for any equations established on a base definition of the form (1.2).

Thus, in any fixed system of units given by the set $L_*, T_*, S_*, \dots$ and relations (1.5), the dimensional analysis is not very useful. It can be used only to find errors in equations, by checking a dimension of its summands.

Additional advantages of using numerical measures in equations are connected with the fact that the values of dimension-dependent quantities can be specified freely. It seems that the ideas of freedom and dependence are mutually exclusive. However, the quantities $\Delta_1, \Delta_2, \dots$ in Eqs. (1.4) can be chosen to be equal to any relative numbers. It is not contradictory to the postulates of the dimensional analysis, and the values of dependent units obtain freedom. This seems unusual because of the tradition of using only fixed systems of units.

For example, there exist problems for which it is convenient to measure lengths in km and areas in m^2. While doing this, $L_* = 1\,\mathrm{km}$ and $S_* = 1\,\mathrm{m}^2$.

In the first equation $s = \Delta_1 \ell^2$ and from (1.3) one can obtain: $\Delta_1 = L_*^2/S_* = 1\,\mathrm{km}^2/1\,\mathrm{m}^2 = 10^6$.

Sometimes it is reasonable to choose $M_* = 1\,\mathrm{kg}$, $W_* = 1\,\mathrm{m/s}^2$, and $F_* = 1\,\mathrm{kG} = 9.81\,\mathrm{n}$. Then Newton's equation, written in terms of numerical measures, takes the form: $f = \Delta_4 mw$, where $\Delta_4 = M_* W_*/F_* = 1/9.81$. The equations based on definitions of area and Newton's Law are changed correspondingly.

It should be noted that the habit of working within the frame of a strictly predetermined system of units, for example, in the International System (SI), becomes a serious psychological obstacle for carrying out transformations of this kind.

2 Introduction of small parameters

2.1 Normalization of equations of motion

In the foregoing it has been established that the units of all the quantities involved in a problem description can be freely chosen. The units that correspond to independent (with respect to dimensions) quantities are free by definition, and the units that correspond to dependent quantities can be changed by setting different values for the right-hand sides of Eqs. (1.4). This fact makes it possible to bring equations to the so-called normalized form.

Consider an arbitrary dynamical system. Equations of its motion written in Cauchy form are:

$$\begin{aligned} \frac{dX_1}{dT} &= F_1(X_1, X_2, \ldots T, A_1, A_2, \ldots, B_1, B_2, \ldots), \\ \frac{dX_2}{dT} &= F_2(X_1, X_2, \ldots T, A_1, A_2, \ldots, B_1, B_2, \ldots), \\ &\ldots, \end{aligned} \tag{2.1}$$

where $X_1, X_2, \ldots X_n$ are phase variables of the problem and $A_1, A_2, \ldots$ and $B_1, B_2, \ldots$ are groups of coefficients with the same dimension.

2.1.1 Procedure of normalization

Normalization of Eqs. (2.1) is fulfilled by some steps, the succession of which can be different.

1. Rewrite the system in terms of numerical measures that correspond to all the quantities involved:

$$T = T_* t,\ X_1 = X_{1*} x_1, \ldots,\ A_1 = A_* a_1, \ldots,\ B_1 = B_* b_1, \ldots. \tag{2.2}$$

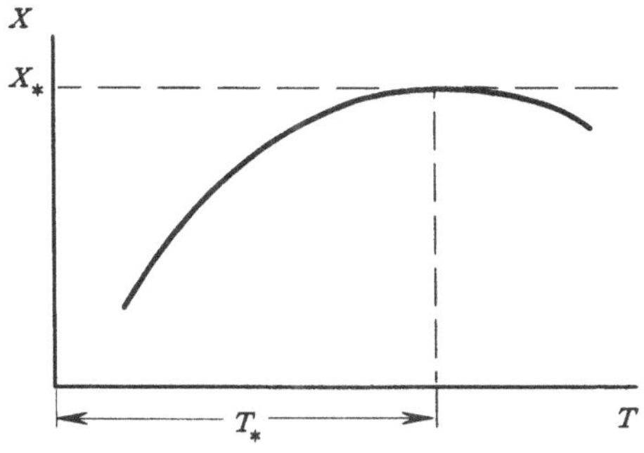

Fig. 1.

2. Outline a class of motions for which the system (2.1) will be considered. Choose $T_*, X_{1*}, \dots, A_*, B_*, \dots$ in (2.2) to be equal to some values that are characteristic for corresponding quantities in this class of motion. The characteristic value for time T_* is determined mainly by the aims of the investigation: by the time-interval on which the system behavior is of interest for a researcher; or on which variables of the problem achieve their limiting values; or by reasons of getting the equations in a more compact form; and so on. An appropriate choice of T_* in the class of motion usually provides a condition:

$$T \leq T_*. \tag{2.3}$$

Characteristic values of phase variables are determined by their maximal absolute values on the time interval given by (2.3) (see Fig. 1). Thus,

$$X_{1*} = \max|X_1|, \qquad X_{2*} = \max|X_2|, \dots . \tag{2.4}$$

By analogy with (2.4), characteristic values of coefficients are assumed to be equal to the maximal absolute values in each group of coefficients:

$$A_* = \max_k\{|A_k|\}, \qquad B_* = \max_h\{|B_h|\}, \dots . \tag{2.5}$$

Using the characteristic values (2.3) – (2.5) as units in (2.2) determines a system of units that is specific for a given class of motion.

In this system of units the absolute values of dimensionless variables $t, x_1, x_2, \dots$ are varied on intervals of the order of unity in accordance with (2.3) and (2.4); and by (2.5), the absolute values of coefficients a_k, b_h are not greater than values of unity order.

3. Divide each equation of (2.1) transformed in accordance with (2.2) – (2.5) by a combination of dimensional multipliers $T_*, X_{1*}, \dots$, which

has the same dimension as the equation. Then (2.1) takes the form:

$$\begin{aligned}
\frac{T_1}{T_*}\frac{dx_1}{dt} &= f_1(x_1, x_2, \ldots, t, \Delta_1, \Delta_2, \ldots), \\
\frac{T_2}{T_*}\frac{dx_2}{dt} &= f_2(x_1, x_2, \ldots, t, \Delta_1, \Delta_2, \ldots), \\
\ldots . &
\end{aligned} \tag{2.6}$$

Equations (2.6) are dimensionless; they are written in terms of dimensionless numerical measures: $t, x_1, \ldots$. Multipliers $T_1/T_*, T_2/T_*, \ldots$ in the left-hand sides of the equations are also dimensionless; therefore $T_1, T_2, \ldots$ have the dimension of time and are often called "time constants" for corresponding variables of the system. Dimensionless groups Δ_1, Δ_2, ... in the right-hand sides are expressed by T_*, $X_{1*}, \ldots$ This completes normalization of Eqs. (2.1).

In the normalized system (2.6) the quantities T_1/T_*, $T_2/T_*, \ldots$, Δ_1, $\Delta_2, \ldots$ correspond to the chosen class of motion. Some of these quantities can be sufficiently small in order to play the role of small parameters in approximate analysis.

It should be noted that normalization of equations is the most important and the less formalized part of fractional analysis of a problem. It depends mainly on the experience and skill of a researcher. To choose proper characteristic values and relations connecting them, one has to use experimental data, analogies, rough estimates of initial equations solution, and the like in correspondence with the aims of the investigation.

It is not necessary to set the characteristic values with a very high accuracy. It must agree with a value of a small parameter and with the required accuracy of approximations. For example, if a small parameter is about 10^{-3}–10^{-4}, the accuracy of characteristic-value setting should be sufficient to be of the order of 10%. This will be proved to provide a 1% error in a zeroth approximation.

2.1.2 Normalization of equations for aperiodic and oscillating motion

1. Aperiodic motion.
Let us consider a Stokes' problem about vertical fall of a ball in a viscous fluid. Let an axis x be directed vertically downwards; then the equations of motion are:

$$\begin{gathered}
M\frac{dV}{dT} = MG - KV, \qquad \frac{dX}{dT} = V; \\
X(0) = X_0, \qquad V(0) = V_0,
\end{gathered} \tag{2.7}$$

where M is a mass of a ball, X is a coordinate of its center, V is a velocity, K is a coefficient of viscous friction, and G is an acceleration of free falling under the action of weight and Archimedean forces.

Equations (2.7) should be normalized. These equations contain six quantities M, V, T, G, K, X that describe the problem. Coefficients M, G, K do not need to be normalized. Normalization of the remaining quantities leads to:

$$t = \frac{T}{T_*}, \qquad v = \frac{V}{V_*}, \qquad x = \frac{X}{X_*}. \tag{2.8}$$

Now it is necessary to estimate characteristic values of the variables. For that, some simplified estimates should be made. From a static-motion equation $MG - KV = 0$, a stationary velocity is $V_1 = MG/K$. From evaluation $MdV/dT = -KV$ it follows that a time constant that characterizes a friction effect on the velocity is $T_1 = M/K$.

After substitution of (2.8) in (2.7) and division of the first and second resulting equations by MG and V_*, respectively, the system (2.7) takes the form:

$$\begin{gathered} \frac{V_*}{GT_*}\frac{dv}{dt} = 1 - \frac{KV_*}{MG}v, \qquad \frac{X_*}{T_*V_*}\frac{dx}{dt} = v; \\ x(0) = \frac{X_0}{X_*}, \qquad v(0) = \frac{V_0}{V_*}. \end{gathered} \tag{2.9}$$

Equations (2.9) contain three dimensionless groups:

$$\Delta_1 = \frac{V_*}{GT_*}, \qquad \Delta_2 = \frac{KV_*}{MG} = \frac{V_*}{V_1}, \qquad \Delta_3 = \frac{X_*}{T_*V_*}. \tag{2.10}$$

They take different values for different classes of motion.

(a) **Class of motion for small values of time $T \ll T_1$ and small initial velocity $V_0 \ll V_1$**

For this class of motion a characteristic velocity has an order of initial velocity, a characteristic acceleration does not depend on friction force, and a characteristic change of coordinate is as for a uniform-speed motion. It is assumed that initial values X_0 are of the same order as the coordinate change.

Let $T_* = \mu T_1$, $V_* = V_0 = \mu V_1$, $X_* = V_* T_*$, where $\mu \ll 1$ is a given number. Dimensionless terms (2.10) take the values $\Delta_1 = 1$, $\Delta_2 = V_*/V_1 = T_*/T_1 = \mu \ll 1$, and $\Delta_3 = 1$. Then Eqs. (2.9) take the form:

$$\begin{gathered} \frac{dv}{dt} = 1 - \mu v, \qquad \frac{dx}{dt} = v, \\ x(0) = \frac{X_0}{V_0 T_*}, \qquad v(0) = 1\,. \end{gathered} \tag{2.11}$$

(b) **Motion on a small time interval $T \ll T_1$ with an arbitrary initial velocity of the same order as the stationary velocity $V \sim V_1$**

Let $T_* = \mu T_1$, $V_* = V_1$, $X_* = V_* T_*$, and $\mu \ll 1$. Then the dimensionless groups (2.10) take the values $\Delta_1 = T_1/T_* = 1/\mu$, $\Delta_2 = 1$, and $\Delta_3 = 1$. Equations (2.9) take the form:

$$\frac{dv}{dt} = \mu(1 - v), \qquad \frac{dx}{dt} = v;$$

$$x(0) = \frac{X_0}{X_*}, \qquad v(0) = \frac{V_0}{V_1}.$$

(c) **Motion on a time interval of the same order as a time constant $T \sim T_1$, with arbitrary initial velocity $V_0 \sim V_1$**

Let $T_* = T_1$, $V_* = V_1$, and $X_* = V_* T_*$. Then $\Delta_1 = 1$, $\Delta_2 = 1$, and $\Delta_3 = 1$. Equations (2.9) can be written in the form:

$$\frac{dv}{dt} = 1 - v, \qquad \frac{dx}{dt} = v;$$

$$x(0) = \frac{X_0}{X_*}, \qquad v(0) = \frac{V_0}{V_1}.$$

(d) **The class of motion on a time interval that is significantly greater than a time constant $T \gg T_1$ with arbitrary initial velocity $V_0 \sim V_1$**

Let $T_*{=}T_1/\mu$, $V_*{=}V_1$, $X_*{=}V_* T_*$, and $\mu \ll 1$. Then $\Delta_1 = T_1/T_* = \mu$, $\Delta_2 = 1$, and $\Delta_3 = 1$. The equations of motion can be written in the form:

$$\mu \frac{dv}{dt} = 1 - v, \qquad \frac{dx}{dt} = v;$$

$$x(0) = \frac{X_0}{X_*}, \qquad v(0) = \frac{V_0}{V_1}.$$

Even this simple example shows how many possibilities exist of forming equations with different structures of parameters included for the description of different classes of motion of one and the same system.

2. Oscillating motion (harmonic oscillator).
Let us consider a motion of a mass point under the action of a viscoelastic force and some perturbation that depends on time explicitly. Motion of this system is described by the equations:

$$M\ddot{X} = -KX - R\dot{X} + F(T/T_{\mathrm{P}});$$
$$X(0) = X_0, \qquad \dot{X}(0) = V_0\,, \tag{2.12}$$

where X is a coordinate, M is a mass of the particle, K and R are coefficients of elasticity and viscosity, F is limited force, and T_{p} is a characteristic value of perturbation: during this time-interval the function $F(T/T_{\mathrm{p}})$ is changed by the value of the order of its maximal absolute value.

Let us rewrite (2.12) in a Cauchy form and fulfill normalization of the variables: $t = T/T_*$, $x = X/X_*$, $v = V/V_*$, and $f = F/F_*$. After dividing the dynamical equation of the system by KX_*, and the kinematical one by V_*, the system can be written:

$$\begin{aligned} &\frac{M}{K}\frac{V_*}{X_*T_*}\frac{dv}{dt} = -\frac{R}{K}\frac{V_*}{X_*}v - x + \frac{F_*}{KX_*}f\left(\frac{T_*}{T_{\mathrm{p}}}t\right), \\ &\frac{X_*}{V_*T_*}\frac{dx}{dt} = v; \qquad x(0) = \frac{X_0}{X_*}, \qquad v(0) = \frac{V_0}{V_*}. \end{aligned} \tag{2.13}$$

The system (2.13) contains dimensionless groups:

$$\begin{aligned} &\Delta_1 = \frac{M}{K}\frac{V_*}{X_*T_*}, \qquad \Delta_2 = \frac{R}{K}\frac{V_*}{X_*}, \\ &\Delta_3 = \frac{F_*}{KX_*}, \qquad \Delta_4 = \frac{T_*}{T_{\mathrm{p}}}, \\ &\Delta_5 = \frac{X_*}{V_*T_*}. \end{aligned} \tag{2.14}$$

Further estimates of these groups are presented for different classes of motion.

(a) **Free oscillations with small friction on a time interval of the order of a free-oscillation period**

To estimate characteristic values of the variables the simplified equation is used, which is obtained from (2.12) for $F \equiv 0,\ R = 0$:

$$\begin{aligned} &M\ddot{X} + KX = 0; \\ &X(0) = X_0, \qquad \dot{X}(0) = V_0. \end{aligned} \tag{2.15}$$

Solution of this equation is:

$$\begin{aligned} &X = A_0 \cos(\omega_0 T + \varphi_0), \\ &\dot{X} \equiv V = -A_0\omega_0 \sin(\omega_0 T + \varphi_0), \end{aligned} \tag{2.16}$$

where $A_0 = \sqrt{X_0^2 + V_0^2/\omega_0^2}$, $\tan\varphi_0 = V_0/X_0\omega_0$, $\omega_0^2 = K/M$. From (2.16), taking (2.4) into account, it follows that

$$\begin{aligned} &X_* = \max|X| = A_0, \\ &V_* = \max|V| = A_0\omega_0. \end{aligned} \tag{2.17}$$

The estimate for V_* also can be obtained without the use of an analytical expression for $V(T)$, but by analyzing a form of the graph $X(T)$. For this, the estimate $V_* = X_*/T_0$ is used, which is valid for monotonous parts of the graph. Here T_0 is a characteristic time of free oscillations, which have an order of a quarter of the period. During this time the variable X is changed by the value of the order X_*. After choosing the value of T_0 to be equal approximately to one-sixth of the period $T_0 = 1/\omega_0$, one can obtain the estimate V_* just as in (2.17). In the following the limited case is considered when initial conditions X_0 and V_0 give an equal contribution to the value of oscillation amplitude A_0. In this case it can be assumed that in (2.17) $X_* = X_0$ and $V_* = X_0\omega_0$. The system behavior is analyzed on a time interval of the order of the oscillation period; consequently it is chosen that $T_* = T_0$.

After substitution of the chosen characteristic values in (2.14), one can obtain: $\Delta_1 = 1$, $\Delta_2 = 2\zeta = R/KT_0$, $\Delta_3 = 0$, and $\Delta_5 = 1$. The system (2.13) takes the form

$$\frac{dv}{dt} = -2\zeta v - x, \qquad \frac{dx}{dt} = v,$$
$$x(0) = 1, \qquad v(0) = \frac{V_0}{X_0\omega_0}. \tag{2.18}$$

In a small friction case in (2.8) one has $\zeta \ll 1$.

(b) **Class of motion with small friction ($\zeta \ll 1$), slow perturbation ($T_0 \ll T_\mathrm{p}$), and on a time interval of the order of an oscillation period ($T \sim T_\mathrm{p}$)**

An equation that is obtained from (2.12) for $R = 0$ and $F = \mathrm{const}$ is taken for an estimatation. It is assumed that $X_0, V_0/\omega_0$ and, in addition, the static deviation $X_\mathrm{S} = F/K$ are the quantities of the same order. Then it also can be assumed that $X_* = X_{\mathrm{S}*} = F_*/K$ and $F_* = \max|F|$. For this normalization from (2.14) it follows that $\Delta_1 = 1$, $\Delta_2 = 2\zeta \ll 1$, $\Delta_3 = 1$, $\Delta_4 = T_0/T_\mathrm{P} \equiv \mu \ll 1$, and $\Delta_5 = 1$. Equations (2.13) take the form

$$\frac{dv}{dt} = -2\zeta v - x + f(\mu t), \qquad \frac{dx}{dt} = v,$$
$$x(0) = 1, \qquad v(0) = \frac{V_0}{X_0\omega_0}.$$

(c) **Motion with small friction ($\zeta \ll 1$) and slow perturbation ($T_0 \ll T_\mathrm{p}$) on a large time interval of the order of characteristic perturbation time ($T \sim T_\mathrm{p}$)**

In the course of normalization the estimates of characteristic values of all the variables remain the same as in the previous case (b), except $T_* = T_p$. After the preceding normalization from (2.14) it follows that $\Delta_1 = \mu \ll 1$, $\Delta_2 = 2\zeta \ll 1$, and $\Delta_3 = \Delta_4 = \Delta_5 = 1$. Equations (2.13) take the form:

$$\mu \frac{dv}{dt} = -2\zeta v - x + f(t), \qquad \mu \frac{dx}{dt} = v,$$

$$x(0) = 1, \qquad v(0) = \frac{V_0}{X_0 \omega_0}.$$

(d) **Classes of motion of the system with significant friction and slow perturbation on large and small time intervals**

A characteristic equation of the system (2.12) is

$$M\lambda^2 + R\lambda + K = 0.$$

Its roots are:

$$\lambda_{1,2} = -\frac{R}{2M} \pm \sqrt{\left(\frac{R}{2M}\right)^2 - \frac{K}{M}}.$$

The case of "big" friction is considered; thus $(R/2M)^2 \gg K/M$. It can be written approximately that $\lambda_1 \approx -R/M$, $\lambda_2 \approx -K/R$, and $|\lambda_1| \gg |\lambda_2|$. Time constants of aperiodic components of the motion are denoted by

$$T_1 = M/R \approx 1/\lambda_1, \qquad T_2 = R/K \approx 1/\lambda_2, \text{ and } T_1 \ll T_2.$$

Then after dividing both sides of (2.12) by K, it can be written:

$$T_1 T_2 \frac{d^2X}{dT^2} = -T_2 \frac{dX}{dT} - X + \frac{1}{K} F\left(\frac{T}{T_p}\right). \tag{2.19}$$

As previously, it is supposed that the values of static deviation for $F = \text{const}$ in (2.19) and initial deviation of the system X_0 are of the same order. It is assumed that $X_* = F_*/K$, where $F_* = \max|F|$. A class of motion is considered for which $T_p \sim T_2$. An estimation of the maximal velocities is fulfilled by taking into account a simplified equation. It is obtained from (2.19) for $T_1 = 0$ and $F = \text{const}$. Then $V_* = \max|dX/dT| = X_*/T_2$.

After the preceding normalization Eq. (2.19) is transformed into the system

$$\begin{gathered} \frac{T_1}{T_*} \frac{dv}{dt} = -v - x + f\left(\frac{T_*}{T_p} t\right), \qquad \frac{T_2}{T_*} \frac{dx}{dt} = v, \\ x(0) = \frac{X_0}{X_*}, \qquad v(0) = \frac{V_0}{V_*}. \end{gathered} \tag{2.20}$$

If the motion on small time intervals $T \sim T_1$ is analyzed, then in (2.20) $T_* = T_1$ should be taken. Thus, Eqs. (2.20) take the form:

$$\frac{dv}{dt} = -v - x + f(\mu\nu t), \qquad \frac{dx}{dt} = \mu v\,,$$

$$x(0) = \frac{X_0}{X_*}, \qquad v(0) = \frac{V_0}{V_*}; \qquad \mu = \frac{T_1}{T_2} \ll 1, \qquad \nu = \frac{T_2}{T_\text{p}} \sim 1.$$

If the motion on a large time interval $T \sim T_2, T_\text{p}$ is analyzed, then $T_* = T_2$ should be chosen. Then from (2.20) it follows that

$$\mu\frac{dv}{dt} = -v - x + f(\nu t), \qquad \frac{dx}{dt} = v,$$

$$x(0) = \frac{X_0}{X_*}, \qquad v(0) = \frac{V_0}{V_*}.$$

2.2 Variants of small parameter introduction

A positive dimensionless quantity that takes values significantly less than unity is called a small parameter. A possibility of introducing small parameters into equations of motion is determined by a priori assumptions about relative smallness of some quantities. In the following situations that often appear in practice, are considered.

2.2.1 Smallness of coefficients

Let us assume that in Eqs. (2.1) in a group of coefficients A_k with the same dimension, the values of the coefficients are significantly different: $|A_i| \ll |A_j|$, $i \neq j$. Their estimates are denoted by $A_{i*} = \max\{|A_i|\}$,

Two Flavors of Small Parameters

$A_{j*} = \max\{|A_j|\}$. It is assumed that $A_* = A_{j*}$. Then in the right-hand sides of normalized equations the quantities a_j and εa_i appear instead of A_j and A_i. It should be noted that $|a_j|, |a_i| \sim 1$ and $\varepsilon = A_{i*}/A_{j*} \ll 1$.

2.2.2 Smallness of time ratios

Let us suppose that in (2.6) all variables except time are normalized and their characteristic values chosen. It is supposed that after this partial normalization the values of time constants are significantly different: $T_1 \ll T_2$.

In the following it is shown that different types of system dependence on a small parameter can be obtained owing to the choice of the characteristic value of time T_* [3, 4, 6]. Such situations appeared in Sec. 2.1.2.

If $T_* = T_1$ is chosen, then it means that a researcher wants to analyze and try to select the processes that occur on a time interval of the order of the smallest time constant T_1. The system (2.6) takes the form

$$\begin{aligned} &\frac{dx_1}{dt} = f_1\,, \\ &\frac{dx_2}{dt} = \mu f_2\,, \quad \mu = \frac{T_1}{T_2} \ll 1, \\ &\dots\,. \end{aligned} \tag{2.21}$$

(Traditionally in literature on this subject, small time ratios are often denoted by μ.) In (2.21) rates of change of the variables x_1 and x_2 are significantly different: $dx_1/dt \sim 1$, $dx_2/dt \sim \mu \ll 1$. In this case x_1 is called the fast variable and x_2 is called the slow variable; it is said that characteristic time variables are strongly different.

While analyzing processes on a time interval of the order T_2, it is chosen that $T_* = T_2$. Then from (2.6) instead of (2.21) the following system can be obtained

$$\begin{aligned} &\mu\frac{dx_1}{dt} = f_1, \quad \mu = \frac{T_1}{T_2} \ll 1, \\ &\frac{dx_2}{dt} = f_2, \\ &\dots\,. \end{aligned} \tag{2.22}$$

In the system (2.22) $dx_1/dt \sim 1/\mu \gg 1$, and $dx_2/dt \sim 1$; therefore, x_1 is again the fast variable and x_2 is the slow one.

2.2.3. Smallness of coordinates

In the theory of oscillations and the theory of control an assumption about "smallness" of the system variables (e.g. because of small initial perturbations) is traditionally used.

After expanding the right-hand parts $F_1, F_2, \dots$ of (2.1) into a Taylor series in its "small" arguments $X_1, X_2, \dots$, Eqs. (2.1) can be written in a matrix form:

$$\frac{dX}{dT} = AX + G(X), \tag{2.23}$$

where A is a matrix of linear-expansion coefficients, and a column $G(X)$ consists of the terms of the second and higher orders of $X_1, X_2, \dots$. The variables $X_1, X_2, \dots$ are dimensional quantities. Generally speaking, they have different dimensions. Therefore, while saying that they are "small," it should be noted with respect to what.

Equations of dynamical systems of concrete types are written in their traditional dimensional units. For example, in different problems it is convenient to measure a length by millimeters, or in meters, or in fractions of the distance between the Earth and the Sun, and so on; the angles are traditionally measured in radians, angular minutes, and the like. Let $[X_1], [X_2], \dots, [T]$ be dimensional units of corresponding variables in the fixed system of dimensions, which is traditional for the concrete system written in (2.23) form.

To introduce small parameters one can assume a priori that $X_1, X_2, \dots$, and, consequently, their characteristic values, $X_{1*}, X_{2*}, \dots$ are small in comparison with their units of dimensions:

$$T_* = [T], \quad X_{1*} = \varepsilon[X_1], \quad X_{2*} = \varepsilon[X_2], \dots, \ \varepsilon \ll 1 . \tag{2.24}$$

(Traditionally, smallness of coordinate ratios is often denoted by ε.) Normalization of the system (2.23) is fulfilled in accordance with the following change of variables:

$$x_1 = \frac{X_1}{X_{1*}}, \quad x_2 = \frac{X_2}{X_{2*}}, \quad \dots, t = \frac{T}{T_*}. \tag{2.25}$$

After substitution of (2.24) and (2.25) in (2.23), the common factor ε is removed. Then each scalar equation of the system is divided (2.23) by the combination $[X_1], [X_2], \dots, [T]$, which has the same dimension as the equation. It should be taken into account that dimensionless groups $[X_1], [X_2], \dots, [T]$, by analogy with (1.5), are equal to unity in a fixed system of units.

Then (2.23) takes the form:

$$\frac{dx}{dt} = ax + \varepsilon g(x), \quad \varepsilon \ll 1 . \tag{2.26}$$

Now in (2.26) small nonlinear terms are formally estimated by the small multiplier ε.

Remark 1. In specific problems not all phase variables may be small, but only some of them.

Remark 2. The value of the small parameter ε and, consequently, the errors of subsequent approximations are determined from (2.24). This fact makes the choice of initial units $[X_1], [X_2], \ldots$ more significant. The system of initial units may be specified by auxiliary normalization fulfilled before the main one (2.24) and (2.25). Such preliminary normalization procedures are fulfilled in the problems described in Sec. 15.1 and 15.2.

2.2.4. Separation of slow motions determined by small nonlinear terms

Now we consider a special case of the system (2.26), namely, the case when in one of the equations, for example, in the second, the right-hand side begins with quadratic terms. Then after normalization in accordance with (2.24) and (2.25) the system (2.23) takes the form:

$$\frac{dx_1}{dt} = a_{11}x_1 + \ldots + a_{1n}x_n + \varepsilon g_1, \qquad \frac{dx_2}{dt} = \varepsilon g_2, \qquad \ldots ,$$

where $dx_1/dt \sim 1$, $dx_2/dt \sim \varepsilon \ll 1$, that is, the small nonlinear right-hand side of the second equation sets and defines slow motion for x_2. Such a situation is usual for gyro theory.

The more general way of separation of slow motions determined by small nonlinear terms is the method of arbitrary constant variation. A linear system, which is obtained from (2.26) for $\varepsilon = 0$, has the form

$$\frac{dx}{dt} = ax. \tag{2.27}$$

Let Φ be a fundamental matrix for (2.27); thus,

$$x = \Phi c \tag{2.28}$$

is the general solution; c is a column of arbitrary constants; and Eqs. (2.28) are considered as a change of variables x by c in (2.26).

After substitution of (2.28) to (2.26), the following equation for the variable c can be obtained.

$$\frac{dc}{dt} = \varepsilon \Phi^{-1} g(\Phi c)\,. \tag{2.29}$$

Equation (2.29) describes slow variation of the "constants" (2.28) due to influence of the small nonlinear terms in (2.26).

2.2.5. Motion decomposition in "stiff" systems

The systems in which some equations contain linear (with respect to phase variables) terms with "large" coefficients, are usually called stiff systems.

Let the right-hand side of one of the equations from the system (2.1) be in the form of

$$KX_j + Q\,, \tag{2.30}$$

where KX_j describes the stiff action with respect to the variable X_j, and the other summands of the right-hand side are denoted by Q. After normalization (2.30) takes the form

$$KX_{j*}x_j + Q_*q\,. \tag{2.31}$$

Let the characteristic values be connected by the relation:

$$KX_{j*} = Q_*\,, \tag{2.32}$$

which reflects equal significance, that is, equal order of the summands of (2.30).

For "finite" values of Q_*, and "large" K, the characteristic value X_{j*} must be "small." Its "smallness" can be estimated by analogy with (2.24) with the help of the following relation

$$X_{j*} = \varepsilon[X_j]\,. \tag{2.33}$$

After normalization in accordance with (2.33) the expression (2.31) takes the form $x_j + q$.

Now the equation of the system (2.1) with the derivative dX_j/dT of the variable under stiff action is normalized. It is assumed that $X_j = X_{j*}x_j$, $T = [T]t$, ..., and (2.33) is taken into account. After dividing the obtained equation by the combination of dimensional units $[X_j]/[T]$ the left-hand side takes the form $\varepsilon dx_j/dt$. This shows that the variable x_j, upon which the stiff action is performed, is the fast variable.

Interdependence of characteristics of accuracy and time is reflected in the terminology: the systems with high accuracy are often called the systems with high performance.

2.3 Regular and singular perturbations with respect to the small parameter

Normalization of the equations and introduction of the small parameters complete the first stage of fractional analysis. In the next stage, smallness of the parameter is used for constructing an approximate solution.

Let the small parameters (ε or μ) be equal to zero. Initial equations containing small parameters are called perturbed equations with respect to small parameters relative to the unperturbed analogies which are obtained for $\varepsilon, \mu = 0$. Thus, to perturbed equations (2.21) and (2.22) correspond their unperturbed analogies

$$\frac{dx_1}{dt} = f_1(x_1, x_2, \ldots, t), \qquad \frac{dx_2}{dt} = 0, \qquad \ldots \tag{2.34}$$

and

$$0 = f_1(x_1, x_2, \ldots, t), \qquad \frac{dx_2}{dt} = f_2(x_1, x_2, \ldots, t), \qquad \ldots\,. \tag{2.35}$$

It is natural to hope that the unperturbed equations are, in the first place, simpler than the initial perturbed ones, and in the second place, that they determine a solution which is close (near) to the solution of the initial equations. The difference between the solutions of perturbed and unperturbed systems is estimated by the Euclidean norm:

$$\Delta = \sqrt{\sum_{i=1}^{n}[x_i(t,\mu) - x_i(t,0)]^2}\,,$$

where n is the order of the system, and $x_i(t,\mu)$ and $x_i(t,0)$ are the components of the solutions of the perturbed and unperturbed systems, respectively.

For this estimate it is necessary to state explicitly the size of the time interval $t \in D$ on which it is valid. Three cases are considered:

(1) finite time interval $0 \leq t \leq t' < \infty$;

(2) asymptotically large time interval $0 \leq t \leq t'/\mu,\ \ \mu \to 0$; and

(3) infinite time interval $0 \leq t < \infty$.

For estimation there exist two cases [7].

1. For $\mu \to 0$ the proximity of the solutions is provided on the whole interval $t \in D$:

$$\sup_D \Delta \to 0 \quad \text{as } \mu \to 0\,. \tag{2.36}$$

   Then the system with a small parameter is called regularly perturbed.

   The problem of this type is determined by Eqs. (2.21) and (2.34) if the estimation is fulfilled on the finite time interval.

2. If the condition (2.36) is not satisfied, then the system with a small parameter is called singularly perturbed.

   For example, the problem of this type is determined by Eqs. (2.21) and (2.34) if the estimation is fulfilled on an asymptotically large time interval. In fact, from unperturbed equations (2.34) it follows that $x_2 = x_2(0) = \text{const}$. On the other hand, in the general case on this interval the solution of a perturbed (with respect to x_2) system drifts from the initial value by a finite quantity.

   The problem determined by Eqs. (2.22) and (2.35) on each of the previously described time intervals is the singular one. In this case the solution of the system (2.35) develops in the space of less dimensionality than that of the system (2.22). The phase trajectories of the system (2.35) lie on the surface determined by the first finite equation $f_1 = 0$ of this system. The initial point of perturbed system (2.22) does not belong to this surface in the general case. Therefore, at the initial moment and at its proximity, the condition of the solutions' (2.22) and (2.35) proximity is not satisfied.

It should be noted that Eqs. (2.34) describe only the motion corresponding to the fast variable x_1. The slow variable x_2 for small characteristic values of time $T_* = T_1$ does not change in the zeroth approximation.

Conversely, Eqs. (2.35) written for significant (large) values of the characteristic time constant $T_* = T_2$ determine only the motion corresponding to the slow variable x_2. Equations for fast variables in this case are transformed into static relations.

Thus, the system takes either regular or singular form depending on the choice of T_*, and fast or slow components of the motion are selected for approximate analysis. One can say that the researcher observes the

dynamical system with the help of a "microscope" or "telescope." The use of different scales of measure shares either small or large details of the phenomenon. One can say also that this procedure gives a formal method for constructing simplified mathematical models of the phenomenon and for analysis of their properties.)

It is obvious that the analogous scale transformations for separation of the small and large details or fractions of the phenomenon can be fulfilled not only with respect to time, but also for other variables of the problem.

2.4 Two types of power series expansion with respect to a small parameter

It is assumed that the system perturbed with respect to a small parameter satisfies requirements of corresponding theorems [8], and its solution continuously depends on time and a small parameter: $x = x(t, \mu)$. It is also assumed that the system is so complicated that we have not succeeded in finding an analytical solution, and we are seeking an approximate solution in terms of a power series expansion with respect to μ:

$$x^{(0)}(t) + \mu x^{(1)}(t) + \mu^2 x^{(2)}(t) + \ldots = \sum_{k=0}^{\infty} \mu^k x^{(k)}(t) . \tag{2.37}$$

Here and in the following $x^{(k)}(t)$ denotes the vectorial coefficient of power series expansion.

A partial sum of the series (2.37) is considered:

$$x_{(N)}(t, \mu) = \sum_{k=0}^{N} \mu^k x^{(k)}(t) , \tag{2.38}$$

and the error $|x(t, \mu) - x_{(N)}||$ of the function $x(t, \mu)$ approximation (2.38) is estimated. Here $||\ldots||$ denotes a norm of a vector.

For estimation there exist two cases [7, 9].

- If $||x - x_{(N)}|| \to 0$ as $N \to \infty$, $\mu = \text{const}$, then the series (2.37) converges in one or another sense to the function $x(t, \mu)$. If the series converges for $\mu \le \mu_0$, then μ_0 is a radius of convergence. If $N = \text{const}$ and $\mu \le \mu_0$, then the estimate of approximation error is

$$||x - x_{(N)}|| = C\mu^{N+1}, \quad C = \text{const} . \tag{2.39}$$

- If $||x - x_{(N)}|| \to 0$ as $\mu \to 0$, $N = \text{const}$, and the error of approximation is determined by the estimate of the same form (2.39) then the series (2.37) approximates the function $x(t, \mu)$ asymptotically (is asymptotic to the function $x(t, \mu)$). It is obvious that a convergent power series (always) possesses asymptotic properties. Asymptotic series for small finite values of μ may be divergent as $N \to \infty$.

The asymptotic approximations are determined for $\mu \to 0$. Therefore, while solving concrete problems with the help of asymptotic methods a researcher cannot avoid heavy doubt. This doubt is provoked by the fact that in any concrete problem a small parameter has a concrete fixed numerical value, and correctness of asymptotic methods becomes ambiguous [10].

An extremely pedantic researcher believes that he or she has a right to use the results obtained by asymptotic methods only in asymptotic sense, that is, for $\mu \to 0$. This attitude makes the reasoning about classes of motion and normalization of equations devoid of sense and, finally, leads to scholastization of the research. In spite of their formal illegality, asymptotic methods are widely used in practical applied investigations. To justify this fact, the following reasons can be listed.

1. For some convenient approximate methods only asymptotic, that is, weaker, properties of the obtained series were proved.

2. The methods are known, for which it is proved that a series of the form (2.22) converges. However, while solving the practical problems, in most cases a researcher does not succeed in constructing all the terms of a series. Usually the researcher has to be content with a short segment of the series consisting of one or two terms. Further calculations become too tedious. In addition, estimation of the series radius of convergence μ_0 is also a nontrivial task and a researcher often does not succeed in solving it. In this case the error of an actually constructed approximation can be estimated only asymptotically.

   And, finally, the main reason:

3. Practical application of asymptotic expansions for solving concrete problems shows that they provide acceptable accuracy which can be

estimated by the formula (2.39) for finite values of μ. These optimistic conclusions are obtained by comparison of asymptotic approximations with the exact solutions of testing problems or with the results of computer calculations.

The author's experience provided that a parameter can be assumed to be small if its value does not exceed 1/3. In this case the error of zeroth approximation does not exceed 30% on the corresponding time interval.

To conclude these reasons we cite Zharkov [11]: "In this complicated field, as it often happens in geophysics, when we say 'yes' it means at the best 'may be.' On the other hand, the 'iron' law of nature often comes to our assistance, according to which it can be assumed that three is an infinitely large quantity with respect to unity, and 1/3 is infinitely small; that is, equals to zero."

2.5 Redundancy in methods of approximation

Malediction of megadimensionality and polyparameterness lies upon a researcher of dynamical systems. Under these conditions an idea of description redundancy excites the natural psychological reaction of tearing away. However, this idea runs through all the analysis steps of any interesting dynamical system.

As a rule, a set of Lagrangian coordinates of a problem and a spectrum of its motions are overabundant with respect to the goal of the investigation. Additional variables such as reactions of constraints and pseudocoordinates are redundant; a set of dimensional units corresponding to a class of motions considered in Sec. 2.2.1 is also redundant in comparison with the International System. And, finally, the ideas of description redundancy

The Idea of Description Redundancy

and its reasonable completion are also used in algorithms of approximated methods.

In the following this idea is tracked in the most commonly used application procedures of approximate analysis: the Poincaré method of successive approximations, an averaging method, and a method of boundary-layer approximation.

In the aforementioned methods an attempt is made to decompose a vector of problem (2.6) variables x into the "main" part $\overline{x}$ and less important additional part $\widetilde{x}$:

$$x = \overline{x} + \widetilde{x}. \tag{2.40}$$

The additive change of variables (2.40) is excessive: $2n$ components of the vectors $\overline{x}$ and $\widetilde{x}$ are introduced instead of n components of the vector x. This excessiveness makes it possible to complete a definition of variables in (2.40) in different ways depending on the goal of analysis and the type of problem.

Different ways of completing the definition generate different approximate algorithms. Their description is presented in the following.

CHAPTER II

Regularly perturbed systems. Expansions of solutions

3 The Poincaré theorem. The algorithm of expansion

In this section an approximate solution of a regularly perturbed system is constructed on a finite interval of normalized dimensionless time. In the examples considered in Sec. 2, dimensional time $T \sim T_1$ corresponds to this interval.

An arbitrary regularly perturbed system is considered:

$$\begin{aligned} \frac{dx_1}{dt} &= f_1(x_1, \ldots, x_n, t, \varepsilon), \quad x_1(0) = x_{10}, \\ &\cdots \\ \frac{dx_n}{dt} &= f_n(x_1, \ldots, x_n, t, \varepsilon), \quad x_n(0) = x_{n0}; \qquad \varepsilon \ll 1, \end{aligned} \tag{3.1}$$

or in matrix form

$$\frac{dx}{dt} = f(x, t, \varepsilon), \qquad x(0) = x_0; \qquad \varepsilon \ll 1. \tag{3.2}$$

The right-hand sides of Eqs.(3.1) and (3.2) are analytic with respect to their arguments.

An approximate solution of (3.2) is sought in the form of a redundant change of variables (2.40)

$$x = \overline{x} + \widetilde{x}. \tag{3.3}$$

The definition (3.3) is completed in the following way. The components of the solution that do not depend on a small parameter are included in the main part $\overline{x}$, and the small components beginning with the first order with respect to ε are included in the addition $\widetilde{x}$. Then after substitution of (3.3) into (3.2), the right-hand side of (3.2) is expanded in a series with respect to small variables $\widetilde{x}$ and ε:

$$\frac{d\overline{x}}{dt} + \frac{d\widetilde{x}}{dt} = f(\overline{x}, t, 0) + \left[\frac{\partial f_i}{\partial x_j}\right]_0 \widetilde{x} + \left[\frac{\partial f_i}{\partial \varepsilon}\right]_0 \varepsilon + \dots . \tag{3.4}$$

In (3.4) the terms of the second and higher order with respect to $\widetilde{x}, \varepsilon$ are denoted by dots, and multipliers $[\partial f_i/\partial x_j]_0, [\partial f_i/\partial \varepsilon]_0, \dots$ are calculated for $x = \overline{x}(t)$, $\varepsilon = 0$.

For the noted requirements to the system (3.2), the function $\widetilde{x}$ is analytic with respect to a small parameter, and by the predetermination (3.3) its expansion with respect to ε must begin with terms of the first order

$$\widetilde{x}(t, \varepsilon) = \varepsilon \widetilde{x}^{(1)}(t) + \varepsilon^2 \widetilde{x}^{(2)}(t) + \dots . \tag{3.5}$$

Here $\widetilde{x}^{(1)}(t)$, $\widetilde{x}^{(2)}(t), \dots$ are unknown vector-functions of time.

Thus, the solution of the system (3.2) is sought in the form of power series (3.3) and (3.5) of small parameter. The coefficients of the series can be obtained by substituting (3.3) and (3.5) in (3.4) and then equating coefficients of equal powers of ε in the right-hand and left-hand parts of the obtained expression.

Setting the terms corresponding to the zeroth power of ε equal, we obtain the so-called generating system:

$$\frac{d\overline{x}}{dx} = f(\overline{x}, t, 0) . \tag{3.6}$$

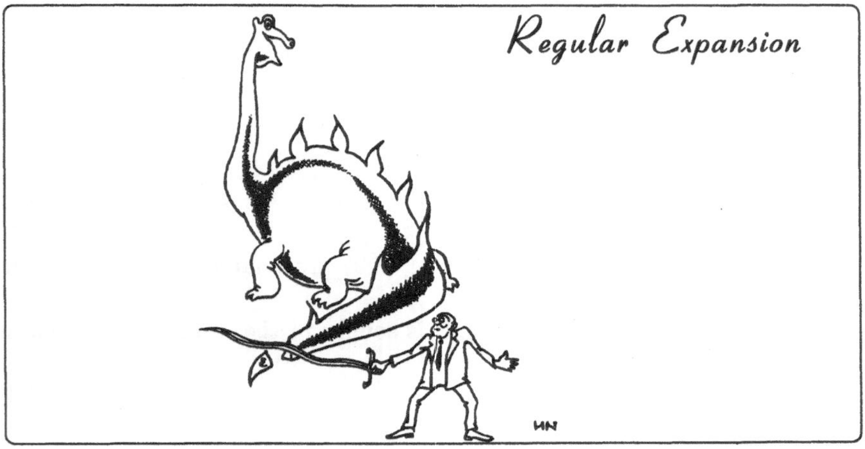

Collecting the terms with the first power of ε leads to the equation:

$$\frac{d\widetilde{x}^{(1)}}{dt} = \left[\frac{\partial f_i}{\partial x_j}\right]_0 \widetilde{x}^{(1)} + f^{(0)}(t). \tag{3.7}$$

Setting the terms with an arbitrary kth power of ε equal leads to the (analogous to (3.7)) equation:

$$\frac{d\widetilde{x}^{(k)}}{dt} = \left[\frac{\partial f_i}{\partial x_j}\right]_0 \widetilde{x}^{(k)} + f^{(k-1)}(t). \tag{3.8}$$

The terms $f^{(0)}(t), f^{(k-1)}(t)$ in (3.7) and (3.8) depend on previous approximations; therefore, they can be assumed to be explicit functions of time.

Initial conditions for Eqs. (3.7) and (3.8) are obtained from (3.3) and (3.5) with $t = 0$:

$$\overline{x}(0) = x_0; \ \widetilde{x}^{(1)}(0) = 0, \ \ldots, \ \widetilde{x}^{(k)}(0) = 0, \ \ldots. \tag{3.9}$$

Equations (3.6) – (3.8) together with (3.9) form an iterative chain of successive approximations to the solution of the system (3.2). It is assumed that the solution of the first in this chain-generating system (3.6) is known.

The Poincaré theorem states that *there exist the finite ε_0, such that if $\varepsilon \le \varepsilon_0$, then the series (3.3) and (3.5) converges uniformly with respect to t to the solution $x(t, \varepsilon)$ of the initial system (3.2) on the finite interval $0 \le t \le t' < \infty$, on which the solution of the generating system is determined.*

Remark 1. Renaming $\overline{x}(t) \equiv x^{(0)}(t)$, $\widetilde{x}^{(k)}(t) \equiv x^{(k)}(t)$ makes it possible to write the series (3.3) and (3.5) in the form:

$$x(t, \varepsilon) = \sum_{k=0}^{\infty} \varepsilon^k x^{(k)}(t). \tag{3.10}$$

Remark 2. The statements of the Poincaré theorem are formulated for the finite time interval. The importance of this restriction can be clarified by the following example. The equation of an oscillator motion with friction is:

$$\frac{d^2x}{dt^2} + \varepsilon\frac{dx}{dt} + \omega_0^2 x = 0; \quad \varepsilon \ll 1. \tag{3.11}$$

The series of the form (3.10) can be constructed for the solution $x(t, \varepsilon)$ of this equation. The term $x^{(0)}(t)$ gives the zeroth approximation of the solution, that is, undamped oscillations with finite amplitude. The error $\Delta^{(0)} = |x(t, \varepsilon) - x^{(0)}(t)|$ of this approximation can be made arbitrarily small on any finite interval of time by choosing a sufficiently small value of ε. This fact corresponds to the estimate of remainder term of convergent (in accordance with Poincaré theorem) series (3.10).

For the system (3.11) with friction, $x(t,\varepsilon) \to 0$ as $t \to \infty$ for any arbitrarily small fixed value of ε. In this case the error $\Delta^{(0)}$ gets finite value, consequently the series (3.10) no longer converges to the solution of the system (3.11).

Remark 3. The preceding restriction on the size of the time interval can be weakened. It is natural that to do this, one has to waive generality of the problem setting.

In Kuz'mina [14] from the very beginning the expansions of the form (3.10) are considered as asymptotic. It is proved that for $\varepsilon \to 0$ the errors of these expansions can be of asymptotic character not only on a finite interval of time, but also on asymptotically large and on infinitely large intervals. The estimates of expansion errors and admissible time intervals are different for different cases of Lyapunov's stability of the first-order approximation of generating system (3.6) solutions.

4 Applications of the Poincaré theorem

4.1 Stokes' problem

Let us consider the problem concerning the vertical fall of a ball in a viscous fluid from Sec. 2.1.2. Poincaré expansion for Eqs.(2.11) obtained by normalization for the class of motions with a small time interval near the initial point has the form:

$$\frac{dv}{dt} = 1 - \mu v, \qquad \frac{dx}{dt} = v; \qquad \mu = \frac{T_*}{T_1} = \frac{V_0}{V_1} \ll 1, \qquad v(0) = v_0 = 1, \quad x(0) = x_0\,. \tag{4.1}$$

The first equation in (4.1) does not depend on the second one. Its solution is sought in accordance with the Poincaré scheme:

$$v = v^{(0)} + \mu v^{(1)} + \mu^2 v^{(2)} + \dots\,. \tag{4.2}$$

Substitution of (4.2) into (4.1) and setting the terms of zeroth order equal leads to

$$\frac{dv^{(0)}}{dt} = 1, \quad v^{(0)}(0) = v_0\,.$$

Then $v^{(0)} = v_0 + t$. The terms of the first order give:

$$\frac{dv^{(1)}}{dt} = -v^{(0)}, \quad v^{(1)} = 0\,.$$

Then

$$v^{(1)} = -v_0 t - \frac{t^2}{2}\,.$$

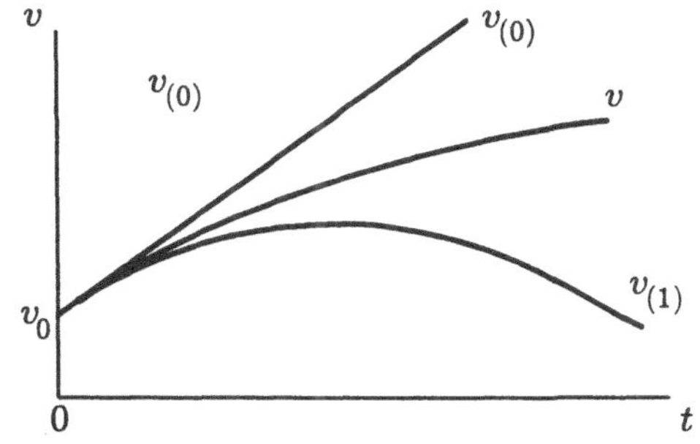

Fig. 2.

It is obvious that

$$v^{(k)} = (-1)^k \left[v_0 \frac{t^k}{k!} + \frac{t^{k+1}}{(k+1)!} \right] .$$

The series obtained can be written in the form:

$$v = v_0 \left(1 - \mu t + \frac{\mu^2 t^2}{2!} - \ldots \right) + \left[-\frac{1}{\mu} + \left(t - \frac{\mu t^2}{2!} + \frac{\mu^2 t^3}{3!} - \ldots \right) \frac{\mu}{\mu} + \frac{1}{\mu} \right] .$$

The series converges to the function

$$v = v_0 e^{-\mu t} + \left(1 - e^{-\mu t}\right) \frac{1}{\mu} ,$$

which is the solution of the initial equation from the system (4.1).

To obtain this solution, it is necessary to construct the whole series. If the construction is limited by one or two terms of the series, then $v_{(0)} = v_0 + t$ and, correspondingly, $v_{(1)} = v_0 + t - \mu(v_0 t + t^2/2)$ has to be taken as the approximate solution.

The graphs of functions $v(t), v_{(0)}(t)$, and $v_{(1)}(t)$ are presented in Fig. 2. It is seen that the approximate solution consisting of a finite number of terms of the series is close to the exact solution only for $\mu t \ll 1$, that is, for $t \ll 1/\mu$. For greater values of time $t \sim 1/\mu$, the value of $\mu t \sim 1$, and the difference between exact and approximate solution is significant. Thus, if a finite number of terms of the series is constructed, then a researcher has to use asymptotic properties of the series, as noted in Sec. 2.4.

Remark. In this example the small parameter was introduced by the assumption about smallness of "time of observation" T_* as compared with time constant T_1. An approximate solution obtained for this assumption is a Taylor series expansion of an exact solution in the proximity of the initial point.

It is obvious that the structure of an approximate solution will be just the same for the system of general kind (2.3), when the characteristic time of observation is significantly less than all partial time constants of the system: $T_* \ll T_1, T_2, \dots$.

In the preceding example we succeeded in constructing the whole Poincaré series, which is impossible for most cases. When researchers deal with finite segments of the series, they are often discouraged by disagreement of the approximate solution and expected results.

4.2 Secular terms

4.2.1 An influence of small variability of stiffness coefficients on the motion of harmonic oscillator

The equation of motion has the form:

$$M\frac{d^2X}{dT^2} + (K_0 + \Delta K)X = 0, \qquad X(0) = X_0, \qquad \frac{dX}{dT}(0) = V_0\,, \tag{4.3}$$

where X is a coordinate, M is the mass of the oscillator, K_0 is the nominal value of the stiffness coefficient, and $\Delta K \ll K_0$ are small deviations from the nominal value.

The motion of the system (4.3) is close to the motion of an oscillating system with small friction from Sec. 2.1.2, case (a). Therefore, both estimation of characteristic values and normalization of Eqs. (4.3) are fulfilled in just the same way. Normalized equations corresponding to Eqs. (4.3) can be written in the form:

$$\frac{dv}{dt} = -(1+\varepsilon)x, \qquad x(0) = 1, \qquad \varepsilon = \frac{\Delta K}{K_0} \ll 1, \qquad \frac{dx}{dt} = v, \qquad v(0) = \frac{V_0}{X_0\omega_0}, \qquad \omega_0^2 = \frac{K_0}{M}\,. \tag{4.4}$$

To simplify calculations, the initial velocity is assumed to be equal to zero ($v(0) = 0$), and the solution of the system (4.4) is sought in the form (3.10):

$$x = x^{(0)} + \varepsilon x^{(1)} + \dots, \qquad v = v^{(0)} + \varepsilon v^{(1)} + \dots\,. \tag{4.5}$$

Equations obtained for the terms of zeroth order with respect to ε are:

$$\frac{dv^{(0)}}{dt} = -x^{(0)}, \qquad \frac{dx^{(0)}}{dt} = v^{(0)}; \qquad x^{(0)}(0) = 1, \quad v^{(0)} = 0\,,$$

or

$$\frac{d^2x^{(0)}}{dt^2} + x^{(0)} = 0; \qquad x^{(0)}(0) = 1, \quad \frac{dx^{(0)}}{dt}(0) = 0\,. \tag{4.6}$$

From (4.6) it follows that

$$x^{(0)} = 1 \cdot \cos t\,. \tag{4.7}$$

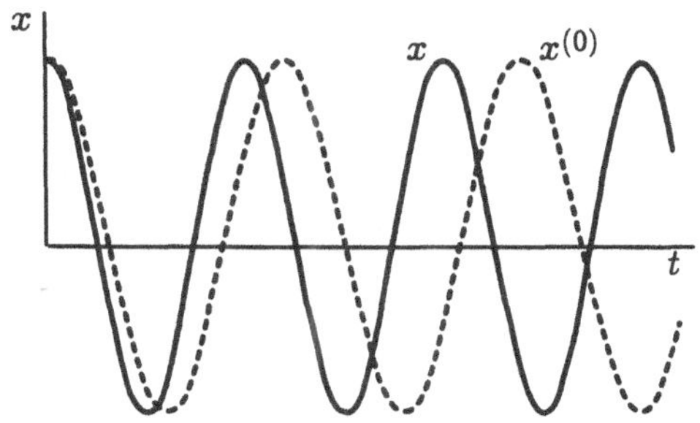

Fig. 3.

From (4.4) and (4.5) for the terms of the first order with respect to ε it follows that

$$\frac{d^2x^{(1)}}{dt^2} + x^{(1)} = -x^{(0)}; \qquad x^{(1)}(0) = 0, \qquad \frac{dx^{(1)}}{dt}(0) = 0. \tag{4.8}$$

The right-hand side of Eq. (4.8) contains a perturbation (4.7) of the resonance frequency.

The corresponding particular solution of (4.8) can be written in the form

$$x^{(1)} = -\frac{t}{2}\sin t. \tag{4.9}$$

Combining (4.5), (4.7), and (4.9) leads to

$$x = x^{(0)} + \varepsilon x^{(1)} + \ldots = \cos t - \varepsilon\frac{t}{2}\sin t + \ldots. \tag{4.10}$$

In (4.10) the summand appeared that grows with time. In approximation theory such components of the solution are called secular terms.

The obtained result cannot help being discouraging, because the solution of the initial system (4.4) has oscillating character and cannot contain components that grow with time.

This illusory contradiction can be explained simply. The graphs of the system (4.4) and (4.6) solutions are presented in Fig. 3. Due to the difference between the frequencies of solutions, the difference between the phases of oscillations of the exact solution $x(t,\varepsilon)$ and the approximate solution $x^{(0)}(t)$ increases with time. Therefore, the difference between two graphs increases with time, and this fact is detected by secular terms (4.9).

This explanation also can be obtained analytically. The exact solution of the initial system (4.4) can be written in the form:

$$x = 1 \cdot \cos \nu t\,, \tag{4.11}$$

where $\nu = \sqrt{1+\varepsilon}$ is the frequency of oscillations. The right-hand side of (4.11) can be expanded into a series in powers of ε:

$$x = \cos t + \varepsilon(-\frac{t}{2}\sin t) + \varepsilon^2(\dots) + \dots . \tag{4.12}$$

It is natural that expressions (4.10) and (4.12) coincide.

4.2.2 Small oscillations of a physical pendulum

The equations of motion have the form:

$$\begin{gathered} \frac{d^2\Phi}{dT^2} + \omega_0^2 \sin\Phi = 0; \\ \Phi(0) = \Phi_0, \qquad \frac{d\Phi}{dT}(0) = 0, \qquad \omega_0^2 = \frac{mg\ell}{I}, \end{gathered} \tag{4.13}$$

where Φ is an angle of deflection from the vertical; m, I, ℓ are a mass, moment of inertia, and the distance between the point of suspension and mass center, respectively. For simplicity the initial velocities are assumed to be equal to zero.

Transformation of Eq. (4.13) into Cauchy form leads to

$$\frac{d\Omega}{dT} = -\omega_0^2 \sin\Phi, \quad \frac{d\Phi}{dT} = \Omega . \tag{4.14}$$

The normalization has the form:

$$\varphi = \frac{\Phi}{\Phi_*}, \quad \omega = \frac{\Omega}{\Omega_*}, \quad t = \frac{T}{T_*} . \tag{4.15}$$

The class of small oscillations of the pendulum is considered. Then $\Phi_* = \Phi_0 \equiv \varepsilon \ll 1$. The estimate Ω_*, T_* follows from the simplified system that is obtained by substitution of Φ instead of $\sin\Phi$ in (4.14). For the linear harmonic oscillator the characteristic values of the variables are estimated in the same way as in Sec. 2.1.2 and Sec. 4.2.1. Therefore,

$$T_* = T_0 = \frac{1}{\omega_0}, \quad \Omega_* = \omega_0 \Phi_* = \varepsilon\omega_0 . \tag{4.16}$$

Substituting (4.15) and (4.16) into (4.14) we obtain for the chosen class of motion

$$\frac{d\omega}{dt} = -\frac{1}{\varepsilon}\sin\varepsilon\varphi, \quad \frac{d\varphi}{dt} = \omega; \quad \varphi(0) = 1, \ \omega(0) = 0 .$$

Expanding the function $\sin\varepsilon\varphi$ into a series with respect to ε, we finally obtain

$$\begin{gathered} \frac{d\omega}{dt} = -\varphi + \mu\frac{\varphi^3}{3!} - \mu^2\frac{\varphi^5}{5!} + \dots , \\ \omega(0) = 0, \qquad \frac{d\varphi}{dt} = \omega, \quad \varphi(0) = 1, \qquad \mu = \varepsilon^2 \ll 1 . \end{gathered} \tag{4.17}$$

A solution of the system (4.17) is sought in the form:

$$\omega = \omega^{(0)} + \mu\omega^{(1)} + \dots, \qquad \varphi = \varphi^{(0)} + \mu\varphi^{(1)} + \dots. \tag{4.18}$$

After substituting (4.18) into (4.17), the equations corresponding to the terms of zeroth order with respect to μ take the form

$$\frac{d\omega^{(0)}}{dt} = -\varphi^{(0)}, \ \frac{d\varphi^{(0)}}{dt} = \omega^{(0)}; \quad \omega^{(0)}(0) = 0, \ \varphi^{(0)}(0) = 1.$$

Therefore, $\varphi^{(0)} = \cos t$, $\omega^{(0)} = -\sin t$.

The equations for the terms of the first order are:

$$\frac{d\omega^{(1)}}{dt} = -\varphi^{(1)} + \frac{(\varphi^{(0)})^3}{3!},$$
$$\frac{d\varphi^{(1)}}{dt} = \omega^{(1)}; \qquad \omega^{(1)}(0) = 0, \quad \varphi^{(1)}(0) = 0. \tag{4.19}$$

After excluding $\omega^{(1)}$ from (4.19) and substituting into (4.19) the explicit expression $\varphi^{(0)}$, we get:

$$\frac{d^2\varphi^{(1)}}{dt^2} + \varphi^{(1)} = \frac{1}{6}\left(\frac{3}{4}\cos t + \frac{1}{4}\cos 3t\right). \tag{4.20}$$

Equation (4.20) is linear and nonhomogeneous. By the principle of superposition a solution corresponding to each summand of the right-hand side is obtained separately. As it was for (4.8), the frequency of the first summand coincides with the frequency of free oscillations of the system (4.20). Hence, the particular solution for $\varphi^{(1)}$, corresponding to the first summand, will include a resonant term increasing with time containing the multiplier $t \sin t$. In the approximate solution $\varphi_{(1)} = \varphi^{(0)} + \mu\varphi^{(1)}$, as in Sec. 4.2.1, there appears a secular term of the form $\mu t \sin t$.

It can be easily shown that oscillations with limited small amplitude correspond to the exact solution and, therefore, the solution does not contain components that grow with time.

After multiplication of both sides of the first equation of (4.14) by $\Omega = d\Phi/dT$ the variables are separated: $\Omega d\Omega = -\omega_0^2 \sin\Phi d\Phi$. By integrating this equation between the limits from the initial to an arbitrary moment an energy integral is obtained:

$$\frac{\Omega^2}{2} - \omega_0^2 \cos\Phi = \frac{\Omega_0^2}{2} - \omega_0^2 \cos\Phi_0. \tag{4.21}$$

The equation of phase trajectories (4.21) in the proximity of the point $\Phi = 0$, $\Omega = 0$ determines a set of closed concentric curves, which corresponds to periodic motion.

This illusory contradiction is explained in the same way as in Sec. 4.2.1. Oscillations of a physical pendulum are not isochronous; that is, a period of oscillation for the exact solution of (4.14) depends on amplitude of oscillations. Oscillations corresponding to zero with respect to μ linear approximation are isochronous. Due to the difference of periods, the difference between oscillation phases grows with time; therefore, the difference between the solutions also grows with time. This difference is detected by secular terms in the next after zeroth approximation $\varphi_{(1)} = \varphi^{(0)} + \mu\varphi^{(1)}$.

Thus, the appearance of secular terms in the problems from Sec. 4.2.1 and 4.2.2 does not indicate a contradiction. In this case, just as in Sec. 4.2.1, the approximate solution can be used only as an asymptotic approximation on a limited time interval.

4.3 Systematic drifts of a gyro in gimbals. Method of successive approximations.

Equations of the gyro [15] have the form:

$$
\begin{aligned}
&J(\Phi_2)\frac{d^2\Phi_1}{dT^2} + \left(\frac{\partial J}{\partial \Phi_2}\right)\frac{d\Phi_1}{dT}\frac{d\Phi_2}{dT} + H\cos\Phi_2\frac{d\Phi_2}{dT} = 0\,,\\
&B_0\frac{d^2\Phi_2}{dT^2} - \frac{1}{2}\left(\frac{\partial J}{\partial \Phi_2}\right)\left(\frac{d\Phi_1}{dT}\right)^2 - H\cos\Phi_2\frac{d\Phi_1}{dT} = 0\,,\\
&J(\Phi_2) = A_2 + C_1 + (A + A_1 - C_1)\cos^2\Phi_2, \qquad B_0 = A + B_1\,.
\end{aligned}
\tag{4.22}
$$

Traditional notation is used in these equations: $A_2, \ldots, C$ are moments of inertia of rings and rotor, $H = C\Omega_{z_1}$ is the rotor angular momentum of the gyro, and Ω_{z_1} is the absolute angular velocity of the rotor. Traditional gyro theory notation for angles of the ring rotation α, β are changed by Φ_1, Φ_2 to make normalization more convenient.

After introducing new variables

$$
\Omega_1 = \frac{d\Phi_1}{dT}, \quad \Omega_2 = \frac{d\Phi_2}{dT}\,, \tag{4.23}
$$

Eqs.(4.22) can be written in phase variables and normalized in accordance with the formulae:

$$
t = \frac{T}{T_*}, \quad a_2 = \frac{A_2}{I_*}, \ \ldots, \quad \varphi_1 = \frac{\Phi_1}{\Phi_*}, \ \ldots, \quad \omega_1 = \frac{\Omega_1}{\Omega_*}, \ldots\,. \tag{4.24}
$$

The quantities $T_*, I_*, \ldots$ should be estimated. All moments of inertia are assumed to be quantities of the same order, and it is chosen that $I_* = \max\{A_2, \ldots, C\}$. It is also assumed that free oscillations of the gyroscope are caused by an impact with the "finite" impulse of torque ΔM. From the estimate $I_*\Delta\Omega = \Delta M$ the "finite" initial velocity $\Delta\Omega$ due to the impact is obtained. The main assumption is the following: $\Delta\Omega \ll \Omega_{z_1}$.

After linearization of Eqs. (4.22) and (4.23) in the proximity of the initial point $\Phi_2(0)$, the system of the second order with respect to Ω_1, Ω_2 is obtained with an antisymmetric matrix of coefficients of Ω_1, Ω_2. Therefore, at least for a small time interval, the motion with respect to Ω_1, Ω_2 is oscillatory, and the characteristic value of Ω_* has the same order as the initial angular velocity: $\Omega_* = \Delta\Omega$. The estimate of characteristic time T_1 of these nutational oscillations is obtained by setting equal characteristic values of the inertial and gyroscopic terms of (4.22): $I_*\Omega_*/T_1 = H\Omega_*$. Hence, $T_1 = I_*/H = (I_*/C)(1/\Omega_{z_1}) \sim 1/\Omega_{z_1}$ is a quantity of the order of one revolution. The amplitude of nutational oscillations with respect to the angle $\Delta\Phi$ is obtained from the estimate

$$\Delta\Phi = T_1\Omega_* = (I_*/C)(\Delta\Omega/\Omega_{z_1}) \sim (\Delta\Omega/\Omega_{z_1}) \ll 1.$$

The quantity $\Delta\Phi \equiv \varepsilon \ll 1$ is assumed to be a small parameter of the problem. In gyro theory it is usually assumed that $\varepsilon = 1/3440$; then the quantities $\Delta\varphi_1, \Delta\varphi_2$ are measured in angular minutes. The class of nutational motions is chosen for analysis, and it is assumed that

$$T_* = T_1, \quad \Omega_* = \frac{\varepsilon}{T_1}, \quad \Phi_* = 1. \tag{4.25}$$

The last relationship means that the angles of the rings' turning are not small.

Normalization according to (4.24) and (4.25) transforms Eqs. (4.22) and (4.23) into the form:

$$\begin{gathered} j(\varphi_2)\frac{d\omega_1}{dt} + \cos\varphi_2 \cdot \omega_2 = -\varepsilon\Big(\frac{\partial j}{\partial\varphi_2}\Big)\omega_1\omega_2, \\ b_0\frac{d\omega_2}{dt} - \cos\varphi_2 \cdot \omega_1 = \varepsilon\Big(\frac{\partial j}{\partial\varphi_2}\Big)\frac{\omega_1^2}{2}, \\ \frac{d\varphi_1}{dt} = \varepsilon\omega_1, \qquad \frac{d\varphi_2}{dt} = \varepsilon\omega_2. \end{gathered} \tag{4.26}$$

The system (4.26) is regularly perturbed. According to Poincaré, its solution is sought in the form:

$$\varphi_1 = \varphi_1^{(0)} + \varepsilon\varphi_1^{(1)} + \ldots, \ \ldots, \ \omega_1 = \omega_1^{(0)} + \varepsilon\omega_1^{(1)} + \ldots, \ \ldots. \tag{4.27}$$

Substitution of (4.27) into (4.26) and equalizing the terms of zero order with respect to ε yields

$$\begin{gathered} j(\varphi_2^{(0)})\frac{d\omega_1^{(0)}}{dt} + \cos\varphi_2^{(0)} \cdot \omega_2^{(0)} = 0, \\ b_0\frac{d\omega_2^{(0)}}{dt} - \cos\varphi_2^{(0)} \cdot \omega_1^{(0)} = 0, \\ \frac{d\varphi_1^{(0)}}{dt} = 0, \qquad \frac{d\varphi_2^{(0)}}{dt} = 0. \end{gathered} \tag{4.28}$$

As expected, the solutions of (4.28) with respect to $\varphi_1^{(0)}, \varphi_2^{(0)}, \omega_1^{(0)}, \omega_2^{(0)}$, are harmonic functions of nutational dimensionless frequency

$$\nu = \frac{\cos\varphi_2^{(0)}}{\sqrt{j(\varphi_2^{(0)})\cdot b_0}} \sim 1 .$$

The equations of the next approximation are:

$$\begin{aligned} j(\varphi_2^{(0)})\frac{d\omega_1^{(1)}}{UP}dt + \cos\varphi_2^{(0)}\omega_2^{(1)} &= -\left(\frac{\partial j}{\partial\varphi_2}\right)^{(0)}\omega_1^{(0)}\omega_2^{(0)} \\ &\quad -\left(\frac{\partial j}{\partial\varphi_2}\right)^{(0)}\varphi_2^{(1)}\frac{d\omega_1^{(0)}}{dt} + \sin\varphi_2^{(0)}\varphi_2^{(1)}\omega_2^{(0)}, \\ b_0\frac{d\omega_2^{(1)}}{dt} - \cos\varphi_2^{(0)}\cdot\omega_1^{(1)} &= \left(\frac{\partial j}{\partial\varphi_2}\right)^{(0)}\frac{\left(\omega_1^{(0)}\right)^2}{2} - \sin\varphi_2^{(0)}\varphi_2^{(1)}\omega_1^{(0)}, \\ \frac{d\varphi_1^{(1)}}{dt} = \omega_1^{(0)}, &\qquad \frac{d\varphi_2^{(1)}}{dt} = \omega_2^{(0)} . \end{aligned} \tag{4.29}$$

After substituting $\varphi_1^{(0)}, \varphi_2^{(0)}, \omega_1^{(0)}, \omega_2^{(0)}$ into the system (4.29) its right-hand sides become explicit functions of time. The constant components of both right-hand sides are extracted and denoted by $\langle M_1\rangle$ and $\langle M_2\rangle$, respectively. Here $\langle\ldots\rangle$ is a symbol of averaging with respect to time.

The particular solution of the system (4.29) determined by these constant quantities is:

$$\omega_1^{(1)} = \frac{\langle M_1\rangle}{\cos\varphi_2^{(0)}}, \quad \omega_2^{(1)} = -\frac{\langle M_2\rangle}{\cos\varphi_2^{(0)}} .$$

Then from equations of the next approximation it follows that

$$\varphi_1^{(2)} = \omega_1^{(1)}t, \quad \varphi_2^{(2)} = \omega_2^{(1)}t .$$

In the process of calculation it is discovered that $\langle M_1\rangle \neq 0$, $\langle M_2\rangle = 0$. Therefore, a free gyroscope without the action of perturbing torques has a growing with time component for the angle φ_1, which is called Magnus drift [15].

Systematic drifts of gyroscopic devices of more complicated construction are determined analogously [16, 17].

Remark 1. Appearance of a gyroscope motion component growing with time is not connected with the property of the method, that is, with the difference between the periods of oscillations of solutions of the exact and the approximate systems. Appearance of nonperiodical terms of the solution is explained by "integrating" properties of the gyroscope, which formally follows from the existence of zero roots of the characteristic equation for the system (4.28).

Remark 2. The procedure of calculations according to the Poincaré method except for notations repeats the calculations of the method of successive approximations for the systems with small nonlinear terms. Hence, a basis and constraints of the method of successive approximations is established by Poincaré theory.

5 Poincaré–Lyapunov method

5.1 Algorithm of the method

One of the constraints of the Poincaré method is that the approximation can be constructed only for a limited time interval. One can say that limited time is the price for generality of the method which is stated for an arbitrary regularly perturbed system. If the class of dynamical systems under consideration is restricted and contains only systems for which there exist periodic solutions, then the Poincaré method can be transformed and approximations of periodic solutions can be found for the whole time interval. Now we formulate a theorem and present a corresponding scheme of calculations.

A dynamical system

$$\frac{dX_k}{dT} = a_{k1}X_1 + \ldots + a_{kn}X_n + M_k(X_1, \ldots, X_n)\,, \qquad k = 1, \ldots, n \quad (5.1)$$

is analyzed. The system (5.1) has an equilibrium $X_k = 0,\ k = 1, \ldots, n$; M_k are analytic functions, expansions of which in the proximity of equilibrium state begin with the terms of the second order or higher with respect to X_k.

It is assumed that:

1. The characteristic equation for the linear part of (5.1) has two simple roots $\pm i\omega_0$. In this case the linear system corresponding to (5.1) has a periodic solution.

2. There exist no roots of the kind $\pm im\omega_0,\ m = 2, 3, \ldots$; therefore, resonances and "secularities" corresponding to higher harmonics, which can appear due to nonlinear terms of equations of motion, are ruled out.

3. There exist no zero roots, hence the possibility of nonperiodical components of motion such as "Magnus drifts" is ruled out (see Remark 1 in Sec. 4.3).

New variables which are the normal coordinates of the linear part are introduced into (5.1). Therefore, in the transformed system a block of equations appears, which corresponds to the roots $\pm i\omega_0$:

$$\begin{aligned}\frac{dX}{dT} &= -\omega_0 Y + M_X(X,Y,Y_k)\,,\\ \frac{dY}{dT} &= \ \omega_0 X + M_Y(X,Y,Y_k)\,, \qquad (5.2)\\ \frac{dY_k}{dT} &= b_{k1}Y_1 + \ldots + b_{k,n-2}Y_{n-2} + G_k(X,Y,Y_k), \quad k = 1,\ldots,n-2\,.\end{aligned}$$

As previously, in these equations M_X, M_Y, G_k are the higher order of smallness terms. The last assumptions are formulated for the variables of the system (5.2).

4. For the system (5.2) there exists the first integral of the form $H = X^2 + Y^2 + W(X,Y,Y_k) = \text{const}$, where the expansion of W begins with the terms of the third order with respect to X, Y and of the second order with respect to Y_k.

The last assumption is beyond the frame of the linear conditions (1) – (3) and rules out, for example, nonlinear friction due to which periodic motion of the initial nonlinear system would be impossible. Usually for mechanical systems such a first integral is an energy integral as in (4.21) in Sec. 4.2.

If the conditions (1) – (4) are satisfied, then the theorem is valid [18]: *In the proximity of an equilibrium state the system (5.2) has a periodic solution that is analytical with respect to small initial perturbation $X(0)$. A period of this solution T_p is an analytic function with respect to $X(0)$, which tends to the period $2\pi/\omega_0$ of the linear system as $X(0) \to 0$.*

Now the scheme of calculations based on the statements of the theorem can be constructed. Equations (5.2) are normalized according to the expressions:

$$x = \frac{X}{X_*}, \qquad y = \frac{Y}{Y_*}, \qquad y_* = \frac{Y_*}{Y_{k_*}}, \qquad t = \frac{T}{T_*}\,. \qquad (5.3)$$

By analogy with Sec. 2.2.3, it is chosen that $X_* = X(0) = \varepsilon[X]$, and $Y_* = \varepsilon[Y]$, $Y_{k_*} = \varepsilon[Y_k]$. It is chosen that $T_* = T_p/2\pi$ as in Sec. 4.2. An expansion of the period T_p into a series with respect to ε can be written in the form

$$T_p = T_p(\varepsilon) = \frac{2\pi}{\omega_0}(1 + \tau_1\varepsilon + \tau_2\varepsilon^2 + \ldots)\,, \qquad (5.4)$$

where $\tau_1, \tau_2, \ldots$ are unknown coefficients. (We have $T_p \to 2\pi/\omega_0$ as $\varepsilon \to 0$.) Then

$$T_* = \frac{1 + \tau_1\varepsilon + \tau_2\varepsilon^2 + \ldots}{\omega_0}\,. \qquad (5.5)$$

Substitution of (5.3) and (5.5) into (5.2) and cancelling by the factor ε yields

$$\frac{dx}{dt} = [-y + \varepsilon m_x(x, y, y_k, \varepsilon)](1 + \tau_1\varepsilon + \ldots),$$
$$\frac{dy}{dt} = [x + \varepsilon m_y(x, y, y_k, \varepsilon)](1 + \tau_1\varepsilon + \ldots), \tag{5.6}$$
$$\ldots .$$

A solution of the system (5.6) is sought according to the Poincaré theorem:

$$x = x^{(0)} + \varepsilon x^{(1)} + \ldots, \quad y = y^{(0)} + \varepsilon y^{(1)} + \ldots,$$
$$y_k = y_k^{(0)} + \varepsilon y_k^{(1)} + \ldots . \tag{5.7}$$

The constants $\tau_1, \tau_2, \ldots$ are chosen in such a way that the system (5.6) would have a periodic solution in accordance with the statement of the theorem, that is, so that in the solution (5.7) secular terms would not appear.

5.2 Examples. Nonisochronism of nonlinear system oscillations

5.2.1 The problem from Sec. 4.2.1 is considered. As in (5.5) it is assumed that $T_* = (1 + \tau_1\varepsilon + \ldots)/\omega_0$, and additional normalization of time $t = T/T_*$ is fulfilled in the system (4.3). Thus the system (4.3) is transformed into the form (5.6):

$$\frac{dv}{dt} = -(1 + \varepsilon)x(1 + \tau_1\varepsilon + \ldots), \quad \frac{dx}{dt} = v(1 + \tau_1\varepsilon + \ldots),$$

or

$$\frac{d^2x}{dt^2} + (1 + \varepsilon)(1 + \tau_1\varepsilon + \ldots)^2 x = 0. \tag{5.8}$$

The required expansion $x = x^{(0)} + \varepsilon x^{(1)} + \ldots$ is substituted into (5.8). It is natural that an equation coinciding with (4.6) is obtained for terms of the order of ε^0. For terms of the order of ε the following equation is obtained.

$$\frac{d^2x^{(1)}}{dt^2} + x^{(1)} = -(1 + 2\tau_1)x^{(0)}. \tag{5.9}$$

From (5.9) it follows that the solution would not contain secular terms if $\tau_1 = -1/2$. Thus, by (5.4), the first approximation of a period of oscillations for (4.3) is:

$$T_{p(1)} = \frac{2\pi(1 - \frac{\varepsilon}{2})}{\omega_0}. \tag{5.10}$$

The exact value of the oscillation period of the initial system(4.3) is given by the expression:

$$T_{\mathrm{p}} = \frac{2\pi}{\omega_0}\frac{1}{\sqrt{1+\varepsilon}} . \qquad (5.11)$$

The expression (5.10) is obtained by writing the expansion (5.11) with the accuracy of ε.

5.2.2 The problem from Sec. 4.2.2 is considered. The equations of motion of a physical pendulum (4.14) together with the energy integral (4.21) satisfy the conditions of the theorem from Sec. 5. Normalized equations (4.17) are transformed into the form (5.6):

$$\begin{aligned} \frac{d\omega}{dt} &= (-\varphi + \mu\frac{\varphi^3}{3!} + \ldots)(1 + \tau_1\mu + \ldots) , \\ \frac{d\varphi}{dt} &= \omega(1 + \tau_1\mu + \ldots) . \end{aligned} \qquad (5.12)$$

A solution of the system (5.12) is sought in the form (4.18). By analogy with Sec. 4.2.2, $\varphi^{(0)} = \cos t$, $\omega^{(0)} = -\sin t$. The equations for the terms of the first order are

$$\begin{aligned} \frac{d\omega^{(1)}}{dt} &= -\varphi^{(1)} + \frac{1}{6}\left(\varphi^{(0)}\right)^2 - \tau_1\varphi^{(0)} , \\ \frac{d\varphi^{(1)}}{dt} &= \omega^{(1)} + \tau_1\omega^{(0)} . \end{aligned} \qquad (5.13)$$

The variable $\omega_1^{(1)}$ is excluded from (5.3):

$$\frac{d^2\varphi^{(1)}}{dt^2} + \varphi^{(1)} = \tau_1\frac{d}{dt}\omega^{(0)} + \frac{1}{6}(\varphi^{(0)})^3 - \tau_1\varphi^{(0)} ;$$

then $\varphi^{(0)}, \omega^{(0)}$ are substituted:

$$\frac{d^2\varphi^{(1)}}{dt^2} + \varphi^{(1)} = (-2\tau_1 + \frac{1}{8})\cos t + \ldots . \qquad (5.14)$$

Here the inessential third harmonic is denoted by dots.

The condition of absence of secular terms in the solution of Eq. (5.14) is $(-2\tau_1 + 1/8) = 0$; then $\tau_1 = 1/16$. Thus the first approximation $T_{\mathrm{p}(1)}$ of the period of oscillations (5.4) is obtained: $T_{\mathrm{p}(1)} = 2\pi(1 + \tau_1\mu)/\omega_0$. The corresponding frequency is $\omega_{0(1)} = \omega_0(1 - \tau_1\mu)$.

In Sec. 4.2 a small parameter was introduced: $\mu = \varepsilon^2 = \Phi_0^2$. Therefore,

$$\omega_{0(1)} = \omega_0\left(1 - \frac{\Phi_0^2}{16}\right) . \qquad (5.15)$$

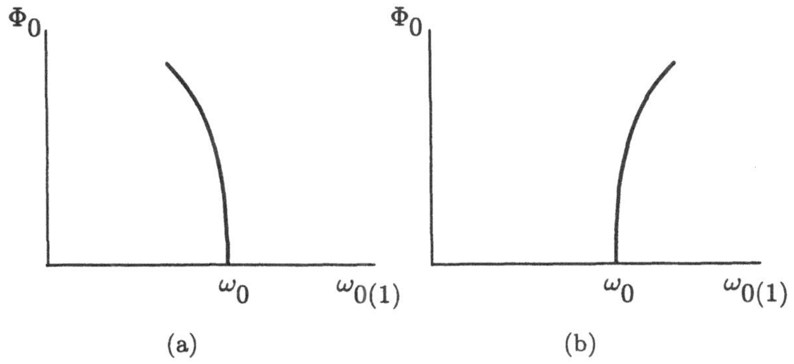

Fig. 4.

In Fig. 4(a) the graph of dependence of the frequency of free oscillations of a physical pendulum on the oscillations' amplitude is shown. Nonisochronism of oscillations is easily seen: the frequency decreases as the amplitude increases.

This type of dependence of the frequency on the amplitude is obtained for oscillating systems with the so-called "soft" characteristic of returning force when the graph of the nonlinear elastic characteristic lies lower than the graph of its linear part.

In the case of the nonlinear "stiff" characteristic, it lies higher than the corresponding linear part. Let nonlinear terms begin with the cubic one. In this case the sign of the term with μ in the expansion of the right-hand side of the first equation in (4.17) changes. Therefore, the sign of the coefficient τ_1, which is determined from the equation of the form of (5.14) changes also. Then dependence of the frequency on the amplitude is presented by Fig. 4(b).

CHAPTER III

Decomposition of motion in systems with fast phase

Among the problems considered in the previous chapter special attention was paid to slowly changing motions parameters, for example, slowly increasing amplitudes in Sec. 4.2, and slow Magnus drifts in Sec. 4.3. The Magnus drifts were obtained by averaging the right-hand side of Eqs. (4.29) with respect to the time variable which is a part of high-frequency oscillating components. During the process of averaging, slow variables φ_1, φ_2 were assumed to be constant and equal to their initial values. This averaging procedure is correct only on bounded intervals of the dimensionless nutational time $t = T/T_1 \sim 1$ where the Poincaré theorem is valid and φ_1, φ_2 changes by values of the order ε according to (4.26). An attempt to use Magnus's formula for large intervals of time $t \sim 1/\varepsilon$ contradicts the theorem. The foregoing procedure of averaging the right-hand sides of (4.29) loses its meaning because during the time interval $t \sim 1/\varepsilon$ the variables φ_1, φ_2 change by finite value.

It is a fruitful idea to separate the slow motion components with the help of averaging the right-hand sides of the equations, and this idea is the foundation of numerous averaging methods. The interval where these methods are valid is extended to $t \sim 1/\varepsilon$. Without this it would be possible to use the regular Poincaré method. The cost for extension of the time interval is an additional restriction on the form of systems being investigated and the fact that approximations to be obtained are asymptotic [13, 19–22].

6 Method of averaging in systems with a single fast phase

6.1 Krylov – Bogolyubov equations in standard form

The Krylov – Bogolyubov equations of motion in standard form are [19]:

$$\frac{dy}{dt} = \varepsilon Y(t, y), y(0) = y_0 , \tag{6.1}$$

where y is the vector of phase variables of arbitrary dimensionality, and $\varepsilon \ll 1$ is a small parameter. All variables in (6.1) are slow. In the following presentation of motion decomposition methods, slow variables are denoted by y and fast variables by z. The functions Y in (6.1) are assumed to be 2π-periodic in time t and to satisfy the requirements of smoothness mentioned in Bogolyubov and Mitropolsky [19].

The system described by (6.1) is regularly perturbed if it is investigated on the finite time interval $t \sim 1$. On this interval it is possible to construct Poincaré approximations of solutions of (6.1) according to Chapter 2, but it is of no interest. On this interval the variable y changes by the value of order ε. To evaluate finite variations of y it is necessary to investigate the system (6.1) on the interval $t \sim 1/\varepsilon$.

The idea of the averaging method is as follows. Due to periodicity properties of the right-hand sides of the system (6.1), its solution is assumed to be a sum of two components: the main slow component $\overline{y}$ and the fast 2π-periodic $\widetilde{y}$ (Fig. 5).

$$y = \overline{y} + \widetilde{y} . \tag{6.2}$$

The formula (6.2) may be considered as a redundant change of variables when two variables $\{\overline{y}, \widetilde{y}\}$ are substituted instead of one variable y. To make use of this redundancy the formula (6.2) should be substituted into (6.1), and average values on the period 2π for both sides of the resulting equation

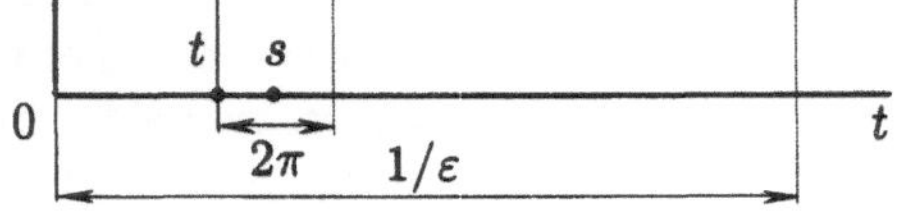

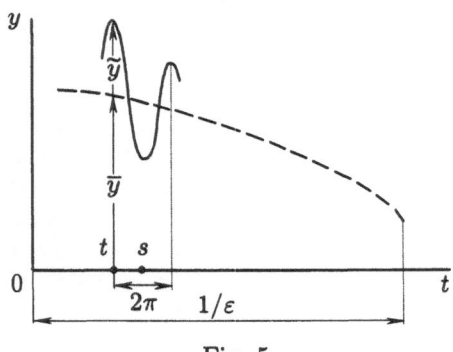

Fig. 5

then should be taken.

$$\frac{1}{2\pi}\int_0^{2\pi}\frac{d}{dt}\Big[\overline{y}(t+s)+\widetilde{y}(t+s)\Big]ds=\varepsilon\frac{1}{2\pi}\int_0^{2\pi}Y\big(t+s,\overline{y}(t+s)+\widetilde{y}(t+s)\big)ds\,. \tag{6.3}$$

Here t is the parameter and s is the variable of integration (Fig. 5). Transposition of operations of differentiation with respect to parameter t and integration for s in the left-hand side of (6.3) yields

$$\frac{d}{dt}\Big[\int_0^{2\pi}(\overline{y}(t+s)+\widetilde{y}(t+s))ds\Big]=\varepsilon\frac{1}{2\pi}\int_0^{2\pi}Y\big(t+s,\overline{y}(t+s)+\widetilde{y}(t+s)\big)ds\,. \tag{6.4}$$

Variations of $\overline{y}$ in (6.4) during one period may be neglected:

$$\overline{y}(t+s)=\overline{y}(t)\,. \tag{6.5}$$

Then

$$\frac{1}{2\pi}\int_0^{2\pi}\overline{y}(t)ds=\overline{y}(t)\,. \tag{6.6}$$

For the functions $\widetilde{y}, Y$ which are periodic with respect to s we have

$$\begin{gathered}\frac{1}{2\pi}\int_0^{2\pi}\widetilde{y}(t+s)ds=0\,,\\ \frac{1}{2\pi}\int_0^{2\pi}Y(t+s,\overline{y}(t)+\widetilde{y}(t+s))ds=Y_0(\overline{y})\,.\end{gathered} \tag{6.7}$$

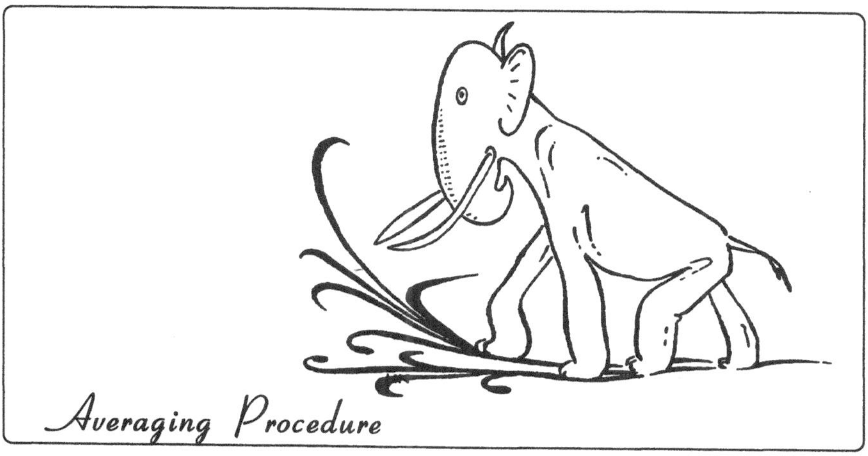

Here $Y_0(\overline{y})$ is the zeroth term of the Fourier-series expansion of the periodic in the s function Y; that is, it is the mean value with respect to s of this function on the period. While calculating Y_0, the value of $\overline{y}$ is assumed to be a parameter. Substitution of (6.5) – (6.7) into (6.4) yields

$$\frac{d}{dt}\overline{y} = \varepsilon Y_0(\overline{y}) . \tag{6.8}$$

The equation (6.8) describes the behavior of the slow component $\overline{y}$, without high-frequency oscillation $\tilde{y}$.

In Bogolyubov and Mitropolsky [19] it is proved that the solution of the averaged system (6.8) is an asymptotic approximation of the solution of the system (6.1) if some conditions imposed on the form of functions Y are satisfied. On a time interval of order $1/\varepsilon$ the approximation error is of order ε.

6.2 Algorithm of asymptotic expansion

The system (6.1) can be written in the form:

$$\begin{aligned} \frac{dy}{dt} &= \varepsilon Y(y, z); \qquad & y(0) &= y_0 , \\ \frac{dz}{dt} &= 1; & z(0) &= z_0 , \end{aligned} \tag{6.9}$$

where y is a slow and z is a fast variable; in averaging methods the latter is usually called the "phase." A rate of change of phase z is called the frequency.

A generalization of (6.9) is given by the system

$$\begin{aligned} \frac{dy}{dt} &= \varepsilon Y(y, z, \varepsilon); \qquad & y(0) &= y_0 , \\ \frac{dz}{dt} &= \omega(y) + \varepsilon Z(y, z, \varepsilon); & z(0) &= z_0 , \end{aligned} \tag{6.10}$$

where, as mentioned previously, y is a vector and the fast phase z is a scalar. It is assumed (traditionally for averaging methods) that Y and Z are 2π-periodic functions in variable z. The right-hand sides of (6.10) are assumed to be analytical functions of their arguments. Then they can be expanded into a series with respect to ε, and Eqs. (6.10) can be written in the form

$$\begin{aligned} \frac{dy}{dt} &= \varepsilon\big[Y^{(0)}(y, z) + \varepsilon Y^{(1)}(y, z) + \ldots\big] , \\ \frac{dz}{dt} &= \omega(y) + \varepsilon Z^{(1)}(y, z) + \varepsilon^2 Z^{(2)}(y, z) + \ldots . \end{aligned} \tag{6.11}$$

In Eqs. (6.11), as opposed to (6.9), the zero-order term $\omega(y)$ of the frequency expansion dz/dt is a slowly changing function of y.

The rule of indexing expansion coefficients in Eqs. (6.11) should be explained. In Sec. 2.1 it is mentioned that one and the same dynamical system (2.6) can be written either in the form (2.21) or in the form (2.22) depending on the choice of time scale T_*. To simplify comparison of the results obtained for different time scales, the indices of expansion coefficients in (6.11) are chosen so that they are independent with respect to the choice of the value T_*.

Now a formal procedure for searching the approximated solution of the system (6.10) can be presented [13, 20].

By analogy with (6.2), the variables in the system (6.10) may be written as the sum of the main components $\overline{y}, \overline{z}$ and small additional components $\widetilde{y}, \widetilde{z}$:

$$y = \overline{y} + \widetilde{y}, \qquad z = \overline{z} + \widetilde{z}\,. \tag{6.12}$$

Expressions (6.12) can be interpreted as the redundant change of variables from $\{y, z\}$ to $\{\overline{y}, \overline{z}, \widetilde{y}, \widetilde{z}\}$. To make this redundancy useful, one can complete a change in (6.12) in a reasonable way:

1. Additional components $\widetilde{y}, \widetilde{z}$ would be the values of order ε as $\varepsilon \to 0$ on the time interval under consideration.

   $$||\widetilde{y}||, |\widetilde{z}| = O(\varepsilon) \qquad \text{as } t \in [0, 1/\varepsilon), \quad \varepsilon \to 0\,, \tag{6.13}$$

   where $O(\varepsilon)$ denotes a value of order ε.

2. The functions $\widetilde{y}$ and $\widetilde{z}$ are assumed to depend on $\overline{y}, \overline{z}$:

   $$\widetilde{y} = \widetilde{y}(\overline{y}, \overline{z}, \varepsilon), \quad \widetilde{z} = \widetilde{z}(\overline{y}, \overline{z}, \varepsilon)\,. \tag{6.14}$$

   It means that, generally speaking, in the process of system evolution, "amplitudes" and "frequencies" of high-frequency additional components in expressions (6.12) are changing slowly as main components $\overline{y}, \overline{z}$ change.

3. It is required that the change of variables (6.12) bring the initial system (6.10) into a simpler form

   $$\begin{aligned} \frac{d\overline{y}}{dt} &= \varepsilon \overline{Y}(\overline{y}, \varepsilon)\,, \\ \frac{d\overline{z}}{dt} &= \overline{\omega}(\overline{y}) + \varepsilon \overline{Z}(\overline{y}, \varepsilon)\,. \end{aligned} \tag{6.15}$$

   By analogy with Eq. (6.8), the first equation in the system (6.15) extracts the main component $\overline{y}$ of the variable y. This equation can be integrated independently. Substitution of its solution into the second equation of (6.15) gives the main component of the system frequency. The variable $\overline{z}$ can be found by integration of this equation.

From (6.15) as $\varepsilon \to 0$ and $t \sim 1/\varepsilon \to 0$ asymptotic evaluation is obtained: $\overline{y} \sim 1$, $\overline{z} \sim 1/\varepsilon \to \infty$. Therefore, the restriction (6.13) can be replaced by the following condition,

$$\|\widetilde{y}\|, |\widetilde{z}| = O(\varepsilon) \qquad \text{as } \overline{z} \to \infty . \tag{6.16}$$

Unknown functions $\widetilde{y}, \widetilde{z}, \overline{Y}, \overline{Z}$ are sought approximately in the form of a series of ε powers. Thus, Eqs. (6.12) through (6.15) take the form:

$$\begin{aligned} y &= \overline{y} + \varepsilon\widetilde{y}^{(1)}(\overline{y},\overline{z}) + \varepsilon^2\widetilde{y}^{(2)}(\overline{y},\overline{z}) + \dots , \\ z &= \overline{z} + \varepsilon\widetilde{z}^{(1)}(\overline{y},\overline{z}) + \varepsilon^2\widetilde{z}^{(2)}(\overline{y},\overline{z}) + \dots , \end{aligned} \tag{6.17}$$

and

$$\begin{aligned} \frac{d\overline{y}}{dt} &= \varepsilon\overline{Y}^{(0)}(\overline{y}) + \varepsilon^2\overline{Y}^{(1)}(\overline{y}) + \dots , \\ \frac{d\overline{z}}{dt} &= \overline{\omega}(\overline{y}) + \varepsilon\overline{Z}^{(1)}(\overline{y}) + \varepsilon^2\overline{Z}^{(2)}(\overline{y}) + \dots . \end{aligned} \tag{6.18}$$

Initial conditions for new variables are obtained by substitution in (6.17) of the value $t = 0$:

$$\begin{aligned} y(0) &= \overline{y}(0) + \varepsilon\widetilde{y}^{(1)}(\overline{y}(0), \overline{z}(0)) + \dots , \\ z(0) &= \overline{z}(0) + \varepsilon\widetilde{z}^{(1)}(\overline{y}(0), \overline{z}(0)) + \dots . \end{aligned}$$

Thus,

$$\begin{aligned} &\overline{y}(0) = y(0), \quad \overline{z}(0) = z(0), \\ &\widetilde{y}^{(k)}(\overline{y}(0), \overline{z}(0)) = \widetilde{z}^{(k)}(\overline{y}(0), \overline{z}(0)) = 0 . \end{aligned} \tag{6.19}$$

After substitution of the expansions (6.17) into (6.11) and taking into account (6.14), one can obtain

$$\begin{aligned} \frac{d}{dt}\widetilde{y}^{(k)} &= \frac{\partial \widetilde{y}^{(k)}}{\partial \overline{y}}\frac{d\overline{y}}{dt} + \frac{\partial \widetilde{y}^{(k)}}{\partial \overline{z}}\frac{d\overline{z}}{dt} , \\ \frac{d}{dt}\widetilde{z}^{(k)} &= \frac{\partial \widetilde{z}^{(k)}}{\partial \overline{y}}\frac{d\overline{y}}{dt} + \frac{\partial \widetilde{z}^{(k)}}{\partial \overline{z}}\frac{d\overline{z}}{dt} . \end{aligned}$$

Collecting terms with $d\overline{y}/dt$, $d\overline{z}/dt$ leads to

$$\begin{aligned} &\Big(E + \varepsilon\frac{\partial \widetilde{y}^{(1)}}{\partial \overline{y}} + \dots\Big)\frac{d\overline{y}}{dt} + \Big(\varepsilon\frac{\partial \widetilde{y}^{(1)}}{\partial \overline{z}} + \dots\Big)\frac{d\overline{z}}{dt} \\ &\qquad\qquad = \varepsilon\big[Y^{(0)} + \varepsilon Y^{(1)} + \dots\big] , \\ &\Big(\varepsilon\frac{\partial \widetilde{z}^{(1)}}{\partial \overline{y}} + \dots\Big)\frac{d\overline{y}}{dt} + \Big(1 + \varepsilon\frac{\partial \widetilde{z}^{(1)}}{\partial \overline{z}} + \dots\Big)\frac{d\overline{z}}{dt} \\ &\qquad\qquad = \omega(\overline{y} + \varepsilon\widetilde{y}^{(1)} + \dots) + \varepsilon Z^{(1)} + \varepsilon^2 Z^{(2)} + \dots . \end{aligned} \tag{6.20}$$

According to Requirement 3 concerning the form of variables change (6.12), the variables $\overline{y}$, $\overline{z}$ must satisfy Eqs. (6.15), that is, Eq. (6.18). Substitution of (6.18) into (6.20) finally produces

$$\begin{aligned} &\left(E+\varepsilon\frac{\partial\widetilde{y}}{\partial\overline{y}}^{(1)}+\ldots\right)\left(\varepsilon\overline{Y}^{(0)}+\ldots\right)+\left(\varepsilon\frac{\partial\widetilde{y}}{\partial\overline{z}}^{(1)}+\ldots\right)\\ &\qquad\times\left(\overline{\omega}(\overline{y})+\varepsilon\overline{Z}^{(1)}+\ldots\right)=\varepsilon\left[Y^{(0)}+\varepsilon Y^{(1)}+\ldots\right],\\ &\left(\varepsilon\frac{\partial\widetilde{z}}{\partial\overline{y}}^{(1)}+\ldots\right)\left(\varepsilon\overline{Y}^{(0)}+\ldots\right)+\left(1+\varepsilon\frac{\partial\widetilde{z}}{\partial\overline{z}}^{(1)}+\ldots\right)\\ &\times\left(\overline{\omega}(\overline{y})+\varepsilon\overline{Z}^{(1)}+\ldots\right)=\omega(\overline{y}+\varepsilon\widetilde{y}^{(1)}+\ldots)+\varepsilon Z^{(1)}+\varepsilon^2 Z^{(2)}+\ldots. \end{aligned} \tag{6.21}$$

Setting the terms with the same powers of ε in (6.21) equal gives:
For ε^0

$$\begin{aligned} &0=0,\\ &\overline{\omega}(\overline{y})=\omega(\overline{y}). \end{aligned} \tag{6.22}$$

For ε^1

$$\begin{aligned} &\overline{Y}^{(0)}(\overline{y})+\frac{\partial\widetilde{y}^{(1)}}{\partial\overline{z}}\overline{\omega}(\overline{y})=Y^{(0)}(\overline{y},\overline{z}),\\ &\overline{Z}^{(1)}(\overline{y})+\frac{\partial\widetilde{z}^{(1)}}{\partial\overline{z}}\overline{\omega}(\overline{y})=\frac{\partial\omega}{\partial\overline{y}}\widetilde{y}^{(1)}+Z^{(1)}(\overline{y},\overline{z}). \end{aligned} \tag{6.23}$$

For ε^k

$$\begin{aligned} &\overline{Y}^{(k-1)}(\overline{y})+\frac{\partial\widetilde{y}^{(k)}}{\partial\overline{z}}\overline{\omega}(\overline{y})=g_{k-1},\\ &\overline{Z}^{(k)}(\overline{y})+\frac{\partial\widetilde{z}^{(k)}}{\partial\overline{z}}\overline{\omega}(\overline{y})=h_{k-1}. \end{aligned} \tag{6.24}$$

The right-hand sides of Eqs. (6.24) are denoted by g_{k-1}, h_{k-1}; they are calculated by analogy with the right-hand sides of Eqs. (6.23).

Equations (6.22) – (6.24) make it possible to obtain Eqs. (6.18) and approximations (6.17). By using Eq. (6.22) and Eq. (6.23) one can conclude that $\overline{\omega}=\omega$ and

$$\widetilde{y}^{(1)}=\frac{1}{\omega(\overline{y})}\int\limits_{z(0)}^{\overline{z}}\left[Y^{(0)}(\overline{y},\overline{z})-\overline{Y}^{(0)}(\overline{y})\right]d\overline{z}+\psi_1(\overline{y}), \tag{6.25}$$

where $\psi_1(\overline{y})$ is an arbitrary function.

The function $\widetilde{y}^{(1)}$ from Eq. (6.25) must satisfy the conditions (6.16). This requirement is fulfilled if the value of $\widetilde{y}^{(1)}$ remains bounded as $\overline{z}\to\infty$.

Since $Y^{(0)}(\overline{y}, \overline{z})$ is a 2π-periodical function in $\overline{z}$, then condition (6.16) is satisfied when

$$\int_0^{2\pi} [Y^{(0)}(\overline{y}, \overline{z}) - \overline{Y}^{(0)}(\overline{y})]\, d\overline{z} = 0,$$

so that

$$\overline{Y}^{(0)}(\overline{y}) = \frac{1}{2\pi} \int_0^{2\pi} Y^{(0)}(\overline{y}, \overline{z})\, d\overline{z}. \tag{6.26}$$

Substitution of Eq. (6.26) into (6.18) and use of initial conditions from (6.19) lead to

$$\frac{d\overline{y}}{dt} = \varepsilon \overline{Y}^{(0)}(\overline{y}), \qquad \overline{y}(0) = y(0). \tag{6.27}$$

Equation (6.27) is similar to Eq. (6.8) and is termed the averaged equation of the first-order approximation.

The function $Y^{(0)}$ in Eq. (6.26) is like Y, the 2π-periodical function in z. Its expansion into the complex form of the Fourier series is:

$$Y^{(0)}(\overline{y}, \overline{z}) = \sum_{|l| \geq 0} Y_l^{(0)} e^{il\overline{z}}, \tag{6.28}$$

where $Y_l^{(0)}(\overline{y})$ are the coefficients of the expansion depending on $\overline{y}$ as on a parameter, and the index l can take on arbitrary integer values, $|l| \geq 0$. After substituting Eq. (6.28) into Eq. (6.26) and averaging over the period, all harmonic terms disappear and

$$\overline{Y}^{(0)}(\overline{y}) = Y_0^{(0)}(\overline{y}), \tag{6.29}$$

which coincides with the zeroth term of the Fourier-series expansion in Eq. (6.28).

The value of y is determined by Eq. (6.27) with the error of order ε, as follows from Eqs. (6.17).

If only the value of y is of interest and ε-order error is assumed to be acceptable, then calculations up to Eq. (6.27) should be finished.

If calculation of phase is also required, then, with the use of the same procedure, from the second equation of (6.23) it is obtained that

$$\overline{Z}^{(1)}(\overline{y}) = \frac{1}{2\pi} \int_0^{2\pi} [Z^{(1)} + \frac{\partial \omega}{\partial \overline{y}} \widetilde{y}^{(1)}]\, d\overline{z}.$$

An especially simple expression is obtained when $\partial\omega/\partial\overline{y} = 0$; that is, ω does not depend on y. Then

$$\overline{Z}^{(1)}(\overline{y}) = \frac{1}{2\pi} \int_0^{2\pi} Z^{(1)}(\overline{y}, \overline{z})\, d\overline{z} \tag{6.30}$$

which is similar to Eq. (6.26).

While solving some problems, computations are often finished at this stage. In this case the approximate system of equations is

$$\begin{aligned} \frac{d\overline{y}}{dt} &= \varepsilon \overline{Y}^{(0)}(\overline{y}); & \overline{y}(0) &= y(0), \\ \frac{d\overline{z}}{dt} &= \omega(\overline{y}) + \varepsilon \overline{Z}^{(1)}(\overline{y}); & \overline{z}(0) &= z(0). \end{aligned} \tag{6.31}$$

If it is necessary to improve the accuracy of the solution, then the computations can be continued. An arbitrary number of the expansion (6.17) terms can be obtained with the help of Eqs. (6.24) complemented by the restrictions (6.16) and the initial conditions (6.19). Thus, if calculation of y with an error ε^2 is required, then it is necessary to find the next term of expansion in Eqs. (6.17). Substitution of Eq. (6.26) into Eq. (6.25) gives the function $\widetilde{y}^{(1)}(\overline{y}, \overline{z})$ with an arbitrary summand $\psi_1(\overline{y})$. However, in accordance with Eqs. (6.19) and (6.25) the condition $\widetilde{y}^{(1)}(\overline{y}(0), \overline{z}(0)) = \psi_1(\overline{y}(0)) = 0$ must be satisfied. Let $\psi_1(\overline{y}(0)) = 0$; then the value of $\widetilde{y}^{(1)}(\overline{y}, \overline{z})$ is determined from Eq. (6.25) uniquely.

6.3 Approximation accuracy

Let us suppose that the terms in the right-hand sides of Eqs. (6.18) up to order ε^N inclusive have been determined after some calculations:

$$\frac{d\overline{y}_{(N)}}{dt} = \varepsilon \sum_{k=1}^{N-1} \varepsilon^k \overline{Y}^{(k)}(\overline{y}_{(N)}), \tag{6.32}$$

where $\overline{y}_{(N)}$ is a solution of this system. Subtraction of Eqs. (6.32) from the first equation of the system (6.18) leads to:

$$\frac{d}{dt}(\overline{y} - \overline{y}_{(N)}) = O(\varepsilon^{N+1}). \tag{6.33}$$

"Residual" $O(\varepsilon^{N+1})$ of the equations has the order ε^{N+1}. Integration of Eq. (6.33) over the time interval $t \sim 1/\varepsilon$ produces an estimate:

$$\overline{y} - \overline{y}_{(N)} = O(\varepsilon^N). \tag{6.34}$$

An analogous estimate for $\overline{z}$ yields

$$\frac{d}{dt}(\overline{z} - \overline{z}_{(N)}) = \omega(\overline{y}) - \omega(\overline{y}_{(N)}) + O(\varepsilon^{N+1}). \tag{6.35}$$

If the function $\omega(\overline{y})$ is continuous, then from Eq. (6.34) it follows that $\omega(\overline{y}) - \omega(\overline{y}_{(N)}) = O(\varepsilon^N)$. Hence,

$$\overline{z} - \overline{z}_{(N)} = O(\varepsilon^{N-1}). \tag{6.36}$$

As is obvious from Eqs. (6.34) and (6.36), on each step of the approximation the fast variable $\overline{z}$ is calculated with accuracy that is by an order of ε lower than that for the slow variable $\overline{y}$.

Expansions (6.17) are also to be constructed with a corresponding accuracy:

$$y_{(N)} = \overline{y}_{(N)} + \varepsilon\tilde{y}^{(1)} + \ldots + \varepsilon^{N-1}\tilde{y}^{N-1)}$$

$$z_{(N-1)} = \overline{z}_{(N-1)} + \varepsilon\tilde{z}^{(1)} + \ldots + \varepsilon^{N-2}\tilde{z}^{N-2)}\,.$$

Then the error of approximation with respect to initial variables is given by the estimates:

$$\Delta y = \|y - y_{(N)}\| = O(\varepsilon^N)$$

$$\Delta z = |z - z_{(N)}| = O(\varepsilon^{N-1})\,.$$

If ω in the initial equation (6.10) does not depend on y, the estimates of accuracy for y and z coincide as follows from Eqs. (6.33) and (6.35). The system (6.31), on which calculations are often finished, gives an error $O(\varepsilon)$ both for y and z; an error of frequency dz/dt is $O(\varepsilon^2)$.

The variable y is determined by the system (6.27) with the error $O(\varepsilon)$. The frequency is calculated from the equation $d\overline{z}/dt = \omega(\overline{y})$ with the error $O(\varepsilon)$. The value of z is determined by this equation for $t \sim 1/\varepsilon$ with finite error.

6.4 Averaging over trajectories of the generating system

The system obtained from (6.10) with $\varepsilon = 0$, which is called a generating system, is given by:

$$\frac{dz}{dt} = \omega(y), \quad z(0) = z_0\,. \tag{6.37}$$

Here the slow variable is assumed to be a parameter that is a consequence of the first equation of the system (6.10). The solution of (6.37) is

$$z = z_0 + \omega(y)t\,. \tag{6.38}$$

The change (6.38) of the integration variable in (6.26) produces

$$\begin{aligned}\overline{Y}^{(0)} &= \frac{1}{2\pi}\int_0^{2\pi} Y^{(0)}(\overline{y}, z_0 + \omega t)dt \\ &= \frac{1}{T_p}\int_0^{T_p} Y^{(0)}(\overline{y}, z_0 + \omega t)dt\,,\end{aligned} \tag{6.39}$$

where $T_p = 2\pi/\omega$ is the period with respect to time t. For the function $Y^{(0)}$ which is periodic in t, its mean value over period (6.39) is equal to the mean value over time:

$$\overline{Y}^{(0)} = \lim_{L\to\infty} \frac{1}{L} \int_0^L Y^{(0)}(\overline{y}, z_0 + \omega t)dt. \tag{6.40}$$

The operation of averaging (6.40) is called "averaging along the trajectories (6.38) of the generating system (6.37)."

After comparison of expressions (6.26), (6.39), and (6.40) it is obvious that all the averaging methods described lead to the same result (6.29).

6.5 Variants of averaging methods

In this section the algorithm of an averaging method has been constructed for the system (6.10), which was considered as the generalization of the system (6.1) in the standard Krylov-Bogolyubov form.

Other methods of presentation are also possible. For example, in Bogolyubov and Mitropolsky [19] and Zhuravlev and Klimov [22] equations of the first approximation of the averaging method are constructed immediately for the standard form

$$\frac{dy}{dt} = \varepsilon Y(y, t, \varepsilon). \tag{6.41}$$

Equations of an arbitrary dynamical system

$$\frac{dx}{dt} = f(x, t) + \varepsilon F(x, t, \varepsilon) \tag{6.42}$$

take the form (6.41) with the help of variable change of the form (2.27)

$$x = g(t, y),$$

where $x = g(t, c)$, $c =$const is the general solution of the unperturbed system, which is obtained from (6.42) with $\varepsilon = 0$.

In Zhuravlev and Klimov [22] the brief proof of the theorem concerning an asymptotic estimate of the error of the averaged first-order approximation equation is presented for a system with periodic and arbitrary function Y depending on time t. Conditions of these estimates' correctness are also presented.

Three statements that are important for solving applied problems [22] should be noted:

1. The requirement of analyticity of equations' (6.10) right-hand sides, mentioned previously, is excessively strong for deducing averaged

equations of first-order approximation. Fulfillment of the condition of piecewise continuity without discontinuity concentration points for the function Y in (6.41) or for the function F in (6.42) is sufficient. The preceding weakening of requirements also makes it possible to solve problems with discontinuous Coulomb friction.

2. Asymptotic estimates for errors of averaged equations of the first-order approximation can be obtained not only for an asymptotically large time interval, but also for a infinite time interval. This extension of time interval is possible if the solution of the averaged system is asymptotically stable. This statement was proved by K. Banfi [22].

3. Stationary solutions of the averaged system correspond to periodic solutions of the initial system if right-hand sides of the initial system are periodic in time. If stationary solutions of the first system are asymptotically stable, then periodic solutions of the second system are also stable [19].

In Zhuravlev and Klimov [22] methods of group theory are briefly presented, which are used for construction of approximations with the same asymptotic properties as approximations obtained by averaging methods.

To obtain equations of the first- and second-order approximations with the help of averaging methods is easy enough. As a rule, further iterations require tedious "hand-made" calculations, which are individual for each concrete system. The methods of group theory make it possible to form the equations of higher approximations using standard recurrent formulae which can be accomplished by computer.

7 Applications of the method of averaging

7.1 Free oscillations with friction of various types

7.1.1 Linear friction

Substituting $F \equiv 0$ into the equation (2.12) of free oscillations of a point mass under action of linear forces of elasticity and friction produces:

$$M\ddot{X} + R\dot{X} + KX = 0, \qquad X(0) = X_0, \qquad \dot{X}(0) = \dot{X}_0. \tag{7.1}$$

After dividing Eq. (1.7) by K, it takes the form:

$$T_0^2\ddot{X} + 2\zeta T_0\dot{X} + X = 0, \tag{7.2}$$

where $T_0^2 = M/K$, $2\zeta T_0 = R/K$.

Substituting $V = dX/dT$ (V for dX/dT) in (7.2) produces a system of equations of the first order. The following normalization is fulfilled,

$$t = \frac{T}{T_*}, \qquad x = \frac{X}{X_*}, \qquad v = \frac{V}{V_*}.$$

The class of motion is considered for which the friction is small: $\zeta \ll 1$. Estimates of characterizing values of T_*, X_*, V_* are assumed to be the same as in Sec. 2.1.2:

$$T_* = T_0 = \frac{1}{\omega_0}, \qquad X_* = X_0, \qquad V_* = \frac{X_*}{T_*}.$$

Then (7.2) takes the form:

$$\begin{aligned} \frac{dv}{dt} &= -x + \varepsilon f, \qquad & v(0) &= \frac{V(0)}{V_*} = \frac{\dot{X}(0)}{X(0)}T_0, \\ \frac{dx}{dt} &= v, & x(0) &= 1, \end{aligned} \tag{7.3}$$

where the following notations are introduced for additional generalization of (7.3)

$$\varepsilon = 2\zeta, \qquad f = -v\,. \tag{7.4}$$

The system (7.4) for $\varepsilon \ll 1$ does not decompose into equations in the fast and slow variables; therefore, the method of averaging cannot be used.

To reduce the system to required form, a change of variables should be used. The following speculations help to obtain an appropriate form of the variable change. For $\varepsilon = 0$ the system (7.3) takes the form:

$$\frac{dv}{dt} = -x, \quad \frac{dx}{dt}v, \quad , x(0) = x_0, \quad v(0) = v_0. \tag{7.5}$$

The solution of (7.5) is:

$$x = a\sin\varphi, \qquad v = a\cos\varphi, \tag{7.6}$$

where a, φ are amplitude and phase of harmonic oscillations that satisfy the equations:

$$\frac{da}{dt} = 0, \quad \frac{d\varphi}{dt} = 1, \quad a(0) = \sqrt{x_0^2 + v_0^2}, \quad \varphi(0) = \arctan\frac{v_0}{x_0}. \tag{7.7}$$

Here and in the following it is assumed that the contributions of the initial conditions $x(0)$, $v(0)$ to initial amplitude $a(0)$ are commensurable; that is, if $x(0) = 1$, then $v(0) \sim 1$.

Relations (7.6) can be interpreted as a change of variables that transfers the system (7.5) with the variables x, v to the system (7.7) with the variables a, φ. In the initial system (7.3) the change of variables (7.6) is fulfilled. It is obvious that additional (as compared with (7.7)) terms of order ε appear in the right-hand sides of the equations for a, φ. Then the variable a is the slow and φ is the fast variable, which was required.

Now a, φ are assumed to be functions of time. After substitution of (7.6) into (7.3), the obtained equations are resolved with respect to da/dt and $d\varphi/dt$:

$$\frac{da}{dt} = f\cos\varphi, \qquad \frac{d\varphi}{dt} = 1 - \frac{\varepsilon}{a} f\sin\varphi. \tag{7.8}$$

Now it is possible to use the averaging method for the investigation of (7.8). For this system approximate equations similar to Eqs. (6.31) have the form:

$$\frac{d\overline{a}}{dt} = \varepsilon\,\overline{Y}^{(0)}, \qquad \frac{\overline{\varphi}}{dt} = 1 + \varepsilon\,\overline{Z}^{(1)}, \tag{7.9}$$

where according to (6.26) and (6.30),

$$\begin{aligned} \overline{Y}^{(0)} &= \frac{1}{2\pi}\int_0^{2\pi} f\cos\varphi\, d\varphi, \\ \overline{Z}^{(1)} &= -\frac{1}{2\pi\,\overline{a}}\int_0^{2\pi} f\sin\varphi\, d\varphi\,. \end{aligned} \tag{7.10}$$

In the problem under consideration in accordance with (7.4) and (7.6) $f = -v = -a\cos\varphi$. Then from (7.10) it follows that

$$\overline{Y}^{(0)} = -\frac{1}{2\pi}\int_0^{2\pi} \overline{a}\cos^2\varphi\, d\varphi = -\frac{\overline{a}}{2},$$

$$\overline{Z}^{(1)} = \frac{1}{2\pi\,\overline{a}}\int_0^{2\pi} \overline{a}\cos\varphi\sin\varphi\, d\varphi = 0\,,$$

and Eqs. (7.9) for $\varepsilon = 2\zeta$ take the form

$$\frac{d\overline{a}}{dt} = -\zeta\,\overline{a}\,, \qquad \frac{d\overline{\varphi}}{dt} = 1.$$

Hence,

$$\overline{a} = a_0 e^{-\zeta t}, \qquad \overline{\varphi} = \varphi_0 + 1 \cdot t\,. \tag{7.11}$$

It is seen that the amplitude of oscillations depends on time exponentially.

It is interesting to compare the approximate solution (7.11) with the exact solution of the initial system (7.3) and (7.4), which can be easily obtained for this linear case. Unit frequency of (7.11) differs from the frequency of oscillations $\sqrt{1-\zeta^2}$ of the system (7.3) and (7.4) by the value of the order ζ^2. The amplitudes and phases of these systems differ for $t \sim 1/\zeta$ by the value of the order ζ. These comparisons are compatible with the general estimates of the method accuracy from Sec. 6.3.

7.1.2 Quadratic friction

A solid moving under the action of a linear force of elasticity and aerodynamic drag is considered. An equation of motion has the form:

$$\begin{aligned} M\ddot{X} &= -KX + F(\dot{X}), \\ X(0) = X_0, &\qquad \dot{X}(0) = \dot{X}_0. \end{aligned} \tag{7.12}$$

In addition to notations (7.1) here $F(\dot{X}) = -(\rho(\dot{X})^2 C_x S \operatorname{sgn} \dot{X})/2$ is aerodynamic drag, ρ is air density, C_x is the dimensionless drag coefficient, and S is the square of the cross section of the solid. The drag is assumed to be small; therefore, it does not disturb the oscillating motion of the solid. Then normalization of (7.12) is fulfilled in the same way as for (7.1), and normalized equations of motion can be written in the form (7.3), where

$$\begin{aligned} \varepsilon &= \frac{\rho C_x S}{K X_0}\left(\frac{X_0}{T_0}\right)^2 = \rho C_x S \frac{X_0}{M}, \\ f &= -v^2 \operatorname{sgn} v = -v|v|. \end{aligned} \tag{7.13}$$

A class of motion is considered for which the condition $\varepsilon \ll 1$ is fulfilled. Consequently, the following approximate procedure is valid only for systems for which a set of "constructional" parameters ρ, C_x, S, M, and a set of initial conditions $X_0, \dot{X}_0$ satisfy this condition.

Calculations presented in the following repeat the scheme of Sec. 7.1.1: with the help of change of variables (7.6), Eqs. (7.3) are transformed to the form (7.8), and then approximate equations (7.9) are obtained.

Let us calculate (7.10) according to (7.6) and (7.3). The limits of integration are changed for convenience.

$$\overline{Y}^{(0)} = \frac{1}{2\pi}\int\limits_{-\pi}^{\pi} [- \overline{a}^2 \cos\varphi|\cos\varphi|] \cos\varphi \, d\varphi.$$

Since the expression under the integral sign is an even function of φ, then

$$\overline{Y}^{(0)} = -\frac{\overline{a}^2}{2\pi} 4 \int\limits_{0}^{\pi/2} \cos^3\varphi \, d\varphi = -\frac{4}{3\pi}\overline{a}^2.$$

By analogy with the preceding, since the function under the integral sign is odd, we obtain

$$\overline{Z}^{(1)} = \frac{1}{2\pi}\int\limits_{-\pi}^{\pi} [- \overline{a} \cos\varphi|\cos\varphi|] \sin\varphi \, d\varphi = 0.$$

Substitution of the expressions $\overline{Y}^{(0)}$, $\overline{Z}^{(1)}$ into approximated equations of the form (7.9) yields:

$$\frac{d\overline{a}}{dt} = -\frac{4\varepsilon}{3\pi}\overline{a}^2, \qquad \frac{d\overline{\varphi}}{dt} = 1;$$

therefore,

$$\overline{a} = \frac{a_0}{1 + (4a_0\varepsilon/3\pi)t}, \qquad \overline{\varphi} = \varphi_0 + 1 \cdot t. \tag{7.14}$$

From (7.14) it is easily seen that the amplitude of oscillations in the case of quadratic friction decreases hyperbolically. Comparison of (7.14) with (7.11) shows that beginning from some moment of time the decrease of the amplitude of oscillations in the case of linear friction becomes faster than in the case of quadratic friction.

7.1.3 Coulomb friction

Let equations of motion be written in the form (7.12), and now

$$F(\dot{X}) = -R_0 \operatorname{sgn} \dot{X}, \qquad R_0 = \text{const}. \tag{7.15}$$

By analogy with Sec. 7.1.1 and 7.1.2, it is assumed that friction does not disturb oscillating motion. Then after the standard normalization described in Sec. 7.1.1, the equations of motion (7.12) and (7.15) take the form (7.3), where

$$\varepsilon = \frac{R_0}{KX_0}, \qquad f = -\operatorname{sgn} v = -\frac{v}{|v|}. \tag{7.16}$$

A class of motion is considered for which the condition $\varepsilon \ll 1$ is satisfied. For this class of motion initial values of the force of elasticity KX_0 are significantly greater than the friction R_0; and in the beginning it accepts getting into the zone of stagnation. For approximate equations (7.9) in accordance with (7.10), (7.6), and (7.16) we have:

$$\begin{aligned} \overline{Y}^{(0)} &= -\frac{1}{2\pi}4\int\limits_{-\pi}^{\pi}\frac{\cos^2\varphi}{|\cos\varphi|}\,d\varphi = -\frac{1}{2\pi}\int\limits_{0}^{\pi/2}\cos\varphi\,d\varphi = -\frac{2}{\pi}, \\ \overline{Z}^{(1)} &= -\frac{1}{2\pi\,\overline{a}}\int\limits_{-\pi}^{\pi}\frac{\cos\varphi}{|\cos\varphi|}\sin\varphi\,d\varphi = 0. \end{aligned} \tag{7.17}$$

Substitution of (7.17) into (7.9) produces

$$\frac{d\overline{a}}{dt} = -\varepsilon\frac{2}{\pi}, \qquad \frac{d\overline{\varphi}}{dt} = 1,$$

then

$$\overline{a} = a_0 - \varepsilon\frac{2}{\pi}t, \qquad \overline{\varphi} = \varphi_0 + 1\cdot t. \tag{7.18}$$

From (7.18) it is seen that an amplitude of oscillations for a system with Coulomb friction decreases as a linear function.

Remark. The problems analyzed in Sec. 7.1.2 and 7.1.3 confirm the assertion that both stages of fractional analysis (small parameters introduction and consequent approximate analysis) are indivisible. Here it is discovered while discussing the condition $\varepsilon \ll 1$, for which the method of small parameter is valid. In the problem described in 7.1.2, the condition $\varepsilon \ll 1$ together with commensurability of the initial conditions $X(0) \sim \dot{X}(0)/T_0$ leads to getting a restriction $(\rho C_x S X_0)/M \ll 1$, in accordance with (7.13). This restriction "forbids" significant initial amplitudes of oscillations. In the problem described in Sec. 7.1.3, the condition $\varepsilon \ll 1$ in accordance with (7.16) "forbids" small initial amplitudes of the order of the stagnation zone. Formal use of the method of small parameter without estimation of the domain of its applicability can produce errors.

7.2 Free oscillations of a tube generator

A scheme of a tube generator is shown in Fig. 6. Current traversing a resistor is denoted by I_R. An equation for I_R (see, e.g., Andronov et al. [8] and Mischenko and Rozov [12]) can be written in the form:

$$L\frac{d^2 I_R}{dT^2} + R\frac{dI_R}{dT} + \frac{1}{C}I_R = \frac{1}{C}f(U), \quad U = M\frac{dI_R}{dT}. \tag{7.19}$$

Here R, C, L are resistance, capacitance and inductance of the circuit, M is

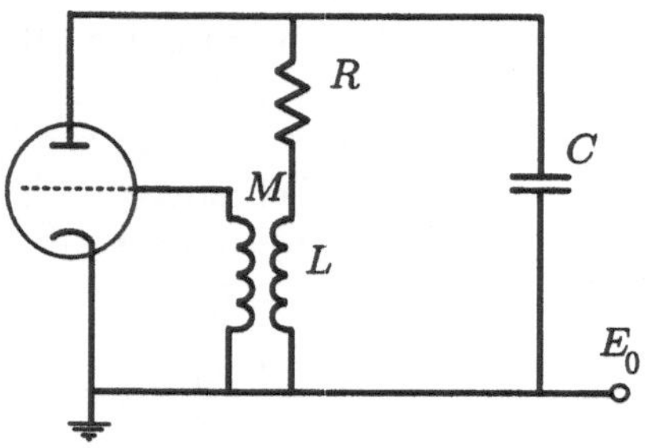

Fig. 6

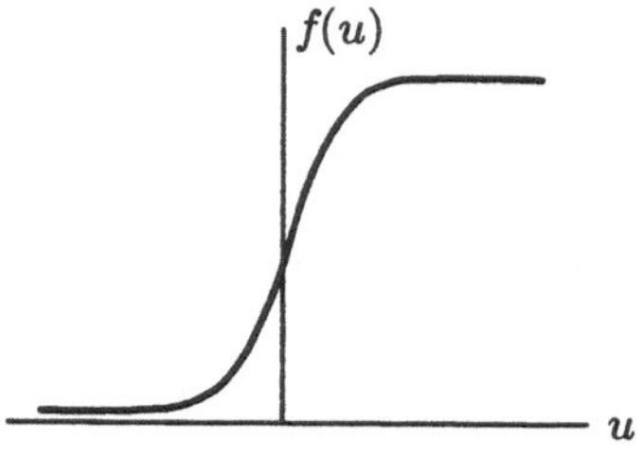

Fig. 7

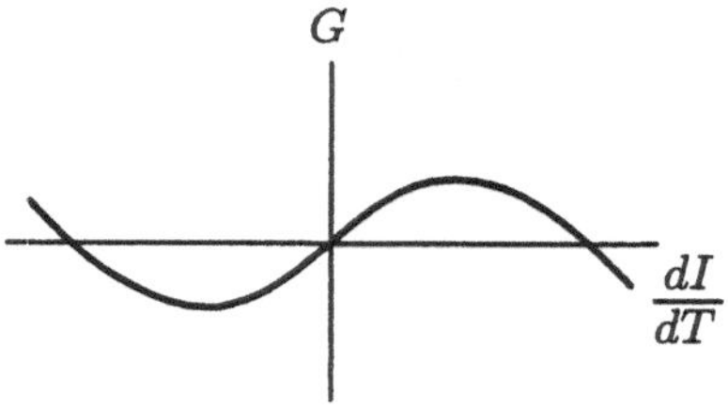

Fig. 8

the coefficient of mutual induction, U is voltage on the grid of a tube, and $f(U)$ is a characteristic of a tube presented in Fig. 7. Substituting $I = I_R - f(0)$ into (7.19) and excluding the variable U yields:

$$LC\frac{d^2T}{dT^2} = G\left(\frac{dI}{dT}\right) - I. \tag{7.20}$$

Here

$$G\frac{dI}{dT} = -RC\frac{dI}{dT} + f\left(M\frac{dI}{dT}\right) - f(0)$$

is a function that determines a "friction" of the system. Its graph is shown in Fig. 8. The system (7.20) has an equilibrium position I=0, dI/dT=0. Stability of this equilibrium position is determined by the value of the derivative of the function $G(dI/dT)$ in zero position: $G'(0)=-RC+f'(0)M$. Here $f'(0)$ is a transconductance of the tube. If $-RC + f'(0)M < 0$, then oscillations of the tube generator are damped; if $-RC + f'(0)M > 0$, then the equilibrium position is unstable. The second case is presented in Fig. 8.

For further discussion it is convenient to approximate $G(dI/dT)$ to a cubic parabolic equation:

$$G\left(\frac{dI}{dT}\right) = [-RC + f'(0)M]\frac{dI}{dT} - A\left(\frac{dI}{dT}\right)^3.$$

Then (7.20) takes the form:

$$LC\frac{d^2I}{dT^2} = [-RC + f'(0)M]\frac{dI}{dT} - A\left(\frac{dI}{dT}\right)^3 - I. \qquad (7.21)$$

After denoting $LC = T_0^2$, $-RC + f'(0) = T_1$, Eq. (7.21) takes the form:

$$T_0^2\frac{d^2I}{dT^2} = \left[T_1 - A\left(\frac{dI}{dT}\right)^2\right]\frac{dI}{dT} - I. \qquad (7.22)$$

In the following only tube generators with the so-called "soft" mode of oscillation [8] are considered. For devices of this kind $T_1 \ll T_0$.

With the help of change of variables $V = dI/dT$, Eq. (7.22) is converted into the system of equations of the first order. To normalize this system, estimates should be done.

Let us analyze the stability of the equilibrium position for (7.22). The equation of the first approximation, which can be obtained by dropping the cubic term with respect to dI/dT in (7.22), shows that the motion for I is oscillating and weakly divergent with close to $1/T_0$ frequency if $T_1 \ll T_0$. From (7.22) it also can be concluded that for "large" values of dI/dT the character of dissipation is determined by the terms $A(dI/dT)^2$. Therefore, we can suppose that for some "mean" amplitude of oscillations the factors that excite oscillations and damp them become balanced. Then the oscillations with frequency of order $1/T_0$ and the amplitude I_* are established in the system. In this case $I \approx I_* \sin(T/T_0)$, $dI/dT = V \approx (I_*/T_0)\cos(T/T_0)$. Therefore, it is assumed for normalization that $T_*{=}T_0$, $V_*{=}I_*/T_0$. Thus, the equation is transferred into a system

$$\frac{dv}{dt} = \frac{T_1}{T_0}\left[1 - A\left(\frac{I_*}{T_0}\right)^2\frac{1}{T_1}v^2\right]v - i, \qquad \frac{di}{dt} = v. \qquad (7.23)$$

As mentioned, in the case of steady oscillations the summands in square brackets should be of the same order. The value of I_* is chosen so that the following equation should be satisfied

$$1 = A\left(\frac{I_*}{T_0}\right)^2\frac{1}{T_1}. \qquad (7.24)$$

Then the system (7.23) can be written in the form (7.3):

$$\frac{dv}{dt} = -i + \varepsilon f, \qquad \frac{di}{dt} = v, \qquad (7.25)$$

where

$$\varepsilon = \frac{T_1}{T_0}, \qquad f = (1 - v^2)v. \qquad (7.26)$$

By analogy with Sec. 7.1, Eqs. (7.26) with the help of change of variables

$$i = a\sin\varphi, \qquad v = a\cos\varphi \qquad (7.27)$$

are transferred to the form

$$\frac{da}{dt} = \varepsilon f \cos\varphi, \qquad \frac{d\varphi}{dt} = 1 - \frac{\varepsilon}{a} f \sin\varphi. \tag{7.28}$$

For the system (7.28) an averaged equation of the first approximation similar to Eq. (6.27) should be constructed.

To demonstrate the use of various methods, in contrast to Sec. 7.1, the right-hand side of (6.27) is calculated with the help of averaging along the trajectories of the generating system. The generating system for (7.28) is the following: $d\varphi/dt = 1$, $\varphi(0) = \varphi_0$. Its solution is:

$$\varphi = \varphi_0 + 1 \cdot t \tag{7.29}$$

Then taking into account Eqs. (6.40), (7.28), (7.27), and (7.29) yields:

$$\begin{aligned} \overline{Y}^{(0)} &= \lim_{L\to\infty} \frac{1}{L} \int_0^L \overline{a}\,[1 - \overline{a}^2 \cos^2(\varphi_0 + t)] \cos^2(\varphi_0 + t)\, dt \\ &= \frac{\overline{a}}{2}(1 - \frac{3}{4}\overline{a}^2), \end{aligned} \tag{7.30}$$

and the required equation similar to (6.27) is:

$$\frac{d\overline{a}}{dt} = \varepsilon \frac{\overline{a}}{2}(1 - \frac{3}{4}\overline{a}^2), \qquad \overline{a}(0) = a_0. \tag{7.31}$$

To analyze Eq. (7.31), stationary solutions $\overline{a}$ = const should be found. Setting the right-hand side of (7.30) equal to zero gives two stationary solutions: $a_1 = 0$, $a_2 = 2/\sqrt{3}$. Their stability should be analyzed. From (7.31) it follows that $da/dt > 0$ on the interval (a_1, a_2) and $da/dt < 0$ on the interval (a_2, ∞). Therefore, a_1 is an unstable stationary solution, which corresponds to the unstable equilibrium point of the initial system (7.25). The solution $a_2 = 2/\sqrt{3}$ is stable, it corresponds to stable harmonic oscillations of the initial system, the so-called self-oscillating mode.

The amplitude of self-oscillations $I_{\max}$ in initial dimensional notations is found with the help of (7.24):

$$I_{\max} = I_* a_2 = 2T_0\sqrt{T_1/3A}.$$

8 Method of harmonic linearization

8.1 Foundations of the method

The solution of problems from Sec. 7.1 and 7.2 demonstrates one of the significant difficulties in using the averaging method. This difficulty is that the system of equations must be previously written in the form (6.10), where the slow and the fast variables are separated. Sometimes this separation is obtained in the course of normalization of initial equations. Sometimes it requires an additional change of variables similar to (7.6) from Sec. 7.1, 7.2, or 2.2. To find the proper change is not a trivial problem. The authors of the most convenient changes are indicated in the literature. Often the form of the change is prompted by the solution structure of the somehow simplified system as mentioned previously. Therefore, the solution is sought in the class of functions approximating this solution one way or another.

Analogous reasoning provides the foundation of the method of harmonic linearization [23]. With the help of this method periodic solutions of nonlinear systems are found. Harmonic functions that a priori approximate sufficiently accurate required periodic solutions are taken as those functions which help construct the solution. The simplicity of harmonic functions stipulates the simplicity of the method, which is one of the most frequently used in the practice of engineering calculations.

The method of harmonic linearization together with all types of averaging methods has its origin in the idea of the integral description of phenomena. Therefore, in Bogolyubov and Mitropolsky [19] a justification of the method of harmonic linearization is accomplished with the help of the averaging method. In Popov and Pal'tov [23] it is shown that the method of harmonic linearization can be reduced to a variant of the averaging method proposed by B.V. Bulgakov [24].

While using the differential method it is not necessary to write the equations of the system in Cauchy form, and normalization of the equations and introduction of small parameter are also unnecessary.

The idea of the method should be clarified for the particular form of the system in which all the equations are linear except one. This equation contains one nonlinear function depending on one variable of the system and, possibly, on its derivative. The system is stationary. (Some other cases are considered in Popov and Pal'tov [23].) The equations of the system have the form:

$$
\begin{aligned}
&A_{11}(D)x_1 + \ldots + A_{1m}(D)x_m = 0, \\
&\ldots\ldots \\
&A_{k1}(D)x_1 + \ldots + A_{km}(D)x_m + f(x_j, Dx_j) = 0, \\
&\ldots\ldots \\
&A_{m1}(D)x_1 + \ldots + A_{mm}(D)x_m = 0.
\end{aligned}
\tag{8.1}
$$

Here $x_1, \ldots, x_j, \ldots, x_m$ are the variables of the system; $A_{11}(D), \ldots$ is the polynomial of the differentiation operator $D = d/dt$ with respect to time t; and $f(x_j, Dx_j)$ is the nonlinear function.

After excluding in (8.1) all the variables except $x_j \equiv x$, it takes the form:

$$Q(D)x = -S(D)f(x, Dx), \tag{8.2}$$

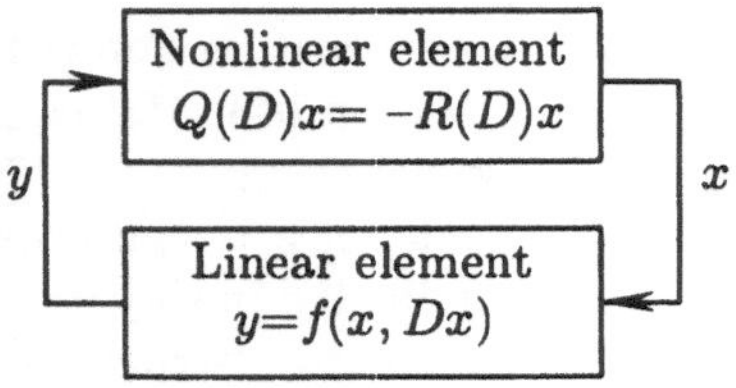

Fig. 9

where $Q(D), S(D)$ are polynomials of D. In Fig. 9 the block-diagram is shown, which corresponds to (8.2).

The first assumption of the method is that the harmonic function of time is assumed to be the input of a nonlinear element in Fig. 9.

$$x = a \sin \Omega t, \tag{8.3}$$

where a and Ω are the unknown amplitude and frequency of the required periodic solution.

Therefore, $Dx = a\Omega \cos \Omega t$, and the nonlinear function

$$f(x, Dx) = f(a \sin \Omega t, a\Omega \cos \Omega t) = f(\varphi) \tag{8.4}$$

is periodic in time t or phase $\varphi = \Omega t$.

The Fourier-series expansion of the periodic function (8.4) has the form:

$$f(\varphi) = A_0 + A_1 \sin \varphi + B_1 \cos \varphi + A_2 \sin 2\varphi + \dots \tag{8.5}$$

In the following only nonlinear functions are considered for which $A_0 = 0$ and expressions for A_1, B_1 can be written in the form:

$$\begin{aligned} A_1(a, \Omega) &= \frac{1}{\pi} \int_0^{2\pi} f(\varphi) \sin \varphi \, d\varphi, \\ B_1(a, \Omega) &= \frac{1}{\pi} \int_0^{2\pi} f(\varphi) \cos \varphi \, d\varphi. \end{aligned} \tag{8.6}$$

The second assumption of the method is that in the expansion (8.5) the harmonics of the second and higher multiplicity with respect to φ are omitted. Then

$$f(x, Dx) = f(\varphi) = A_1 \sin \varphi + B_1 \cos \varphi. \tag{8.7}$$

After taking into account that for periodic motion

$$x = a \sin \Omega t = a \sin \varphi, \qquad Dx = a\Omega \sin \Omega t = a\Omega \cos \varphi,$$

the expression (8.7) can be written in the form

$$f(x, Dx) = qx + q_1 \frac{1}{\Omega} Dx,$$
$$q(a, \Omega) = \frac{A_1}{a}, \qquad q_1(a, \Omega) = \frac{B_1}{a}. \tag{8.8}$$

Thus, for the required periodic motion, the nonlinear function $f(x, Dx)$ is approximated by the linear expression (8.8). This operation is named harmonic linearization.

Now Eq. (8.2), after taking into account (8.8), can be written in the form:

$$Q(D)x = -S(D)\left(qx + q_1 \frac{1}{\Omega} Dx\right). \tag{8.9}$$

This is the linear equation with constant coefficients, depending on unknown variables a, Ω.

To satisfy the first assumption of the method, Eq. (8.9) must have an harmonic solution. The characteristic polynomial for (8.9) with respect to the complex-valued variable p is:

$$L(p) = Q(p) + S(p)\left(q + q_1 \frac{p}{\Omega}\right). \tag{8.10}$$

It must have imaginary root $p = i\omega$, for which the value of ω coincides with the value of required frequency Ω.

Substituting $p = i\omega$ into (8.10) and separating real and imaginary parts of the obtained equation leads to:

$$L(i\omega) = X(\omega, \Omega, a) + iY(\omega, \Omega, a). \tag{8.11}$$

The real and imaginary parts of (8.11) are set equal to zero, and it is assumed that $\omega = \Omega$; then

$$X(\omega, \Omega, a)\big|_{\omega=\Omega} = 0, \qquad Y(\omega, \Omega, a)\big|_{\omega=\Omega} = 0. \tag{8.12}$$

The roots $a = a_0$, $\Omega = \Omega_0$ of the system (8.12) are the amplitude and the frequency of the required solution.

Let us evaluate the condition under which the second assumption of the method about neglecting higher harmonics of the expansion (8.5) is satisfied. As seen in the block diagram in Fig. 9, it is required that while passing the signal $y = f(\varphi) = f(\Omega t)$ through the linear part of the system the higher harmonics of the frequencies $2\Omega_0, 3\Omega_0, \ldots$ are significantly weakened in comparison with the harmonic of the main frequency Ω_0. This condition is called the filter condition [23].

After the periodic solution is found, it is necessary to analyze its stability. If small deviations from the periodic solution decrease with time, then it is asymptotically stable. Periodical motions of this type are called self-oscillations. If small deviations increase, then the unstable periodic solution determines the so-called boundary (bound, limit) of initial conditions, which divides the phase space into the domains with phase trajectories diverging from the periodic solution.

One possible method of stability estimation is connected with the use of the well-known Mikhailov criterion. A hodograph of the function $L(i\omega) = X(\omega, \Omega_0, a_0) + iY(\omega, \Omega_0, a_0)$, is constructed by Eq. (8.11) with $a = a_0$, $\Omega = \Omega_0$. This is a curve tracing the point (X, Y) on the complex plane of these variables for changing the parameter ω from zero to infinity. From (8.12) it follows that the hodograph passes through the origin if the parameter $\omega = \Omega_0$. According to Mikhailov, a hodograph of this kind corresponds to the boundary of the asymptotic stability of a linearized system (8.9).

Let us give a small increment to the amplitude of the found periodic solution $a = a_0 + \Delta a$, and draw a hodograph corresponding to this value of the quantity a. Obviously the hodograph no longer passes through the origin. The origin is placed either "inside" the hodograph, is surrounded by the hodograph, or is "outside." (More accurate definitions of these geometric situations are given in Popov and Pal'tov [23].) The first case, according to the Mikhailov criterion, corresponds to asymptotic stability of the system (8.9), and the second corresponds to instability.

These situations can be distinguished by the relative position of the vector of the tangent to the hodograph on the point $a = a_0, \omega = \Omega_0$ and the vector of the increment $\overline{\Delta L}$ due to the increment Δa. Simple geometrical analysis leads to the following analytical writing of the condition of stability of the periodic solution:

$$\left(\frac{\partial X}{\partial a}\right)_0 \left(\frac{\partial Y}{\partial \omega}\right)_0 - \left(\frac{\partial X}{\partial \omega}\right)_0 \left(\frac{\partial Y}{\partial a}\right)_0 > 0, \tag{8.13}$$

where the lower index after the parentheses means that the values of the derivatives are taken in the point $a = a_0, \quad \omega = \Omega_0$. If the expression in the left-hand side of (8.13) is negative, then the periodic solution is unstable.

Remark. The method of harmonic linearization can be associated with the procedure of function expansion with the use of Chebyshev polynomials.

Let the nonlinear function in (8.2) depend only on the variable x, and be odd.

According to the method of harmonic linearization this function can be approximated by the first harmonic (8.7) of its Fourier-series expansion on the required periodic solution (8.3). Normalization is determined as follows: $\widetilde{x} = x/a$. Then the change of the variable $\widetilde{x}$ in the required solution lies within the interval $[-1, 1]$. Expansion of the function $f(\widetilde{x})$ on this interval with respect to Chebyshev polynomials has the form [25]:

$$f(\widetilde{x}) = A_1 T_1(\widetilde{x}) + A_2 T_2(\widetilde{x}) + \dots .$$

where $T_1(\widetilde{x}) = x$, $T_2(\widetilde{x}), \ldots$ are Chebyshev polynomials, and $A_1, A_2, \ldots$ are the coefficients of the expansion (8.6).

The following approximate equality is assumed to be satisfied: $f(\widetilde{x}) = A_1 T_1(\widetilde{x}) = A_1 \widetilde{x} = A_1 x/a$. An approximation obtained according to Chebyshev coincides with the expression in (8.8) of the method of harmonic linearization.

The preceding comparison of two types of expansions makes it possible to take advantage of useful properties of Chebyshev polynomials, for example, of the method of "telescopic shift" of a power series [25].

The preceding reasoning can be clarified with the help of the Duffing equation: $\ddot{x} + \dot{x} + \varepsilon x^3 = 0$.

Normalization $\widetilde{x} = x/a$ leads to $\ddot{\widetilde{x}} + \widetilde{x} + \varepsilon a^2 \widetilde{x}^3 = 0$; $\widetilde{x} \in [-1, 1]$. The function expansion with the use of Chebyshev polynomials can be written in the form:

$$\widetilde{x}^3 = \frac{3}{4} T_1(\widetilde{x}) + \frac{1}{4} T_3(\widetilde{x}) = \frac{3}{4}\widetilde{x} + \frac{1}{4} T_3(\widetilde{x}); \qquad |T_3(\widetilde{x})| \le 1 .$$

After approximate assumption $\widetilde{x}^3 = 3\widetilde{x}/4$, the Duffing equation is reduced to the linearized according to Chebyshev form:

$$\ddot{\widetilde{x}} + (1 + \frac{3}{4}\varepsilon a^2)\widetilde{x} = 0 .$$

This equation describes harmonic oscillations with an amplitude a, determined by the initial conditions. Dependence of the frequency of free oscillations on the amplitude is represented by the equation:

$$\omega_0 = \sqrt{(1 + \frac{3}{4}\varepsilon a^2} \approx 1 + \frac{3}{8}\varepsilon a^2 ,$$

which coincides with (5.15) at $\varepsilon = \mu/6$ and $a = \Phi_0$.

8.2 Examples.

8.2.1 The problem concerning a tube generator for the "soft" mode of excitation from Sec. 7.2 is considered. After excluding the variable v, Eqs. (7.25) take the form:

$$\frac{d^2 i}{dt^2} - \mu\left[1 - \left(\frac{di}{dt}\right)^2\right]\frac{di}{dt} + i = 0. \tag{8.14}$$

Harmonic linearization of Eq. (8.14) according to (8.8) yields:

$$\frac{d^2 i}{dt^2} - \mu q_1 \frac{di}{dt} + i = 0, \tag{8.15}$$

where $q \equiv 0$, because the nonlinear function $\left[1-(di/dt)^2\right]di/dt$ does not depend on the variable i. Taking into account (8.8) and (8.6) leads to

$$q_1 = \frac{1}{a}B_1, \qquad B_1 = \frac{1}{\pi}\int\limits_0^{2\pi}(1 - a^2\cos^2\varphi)a\cos^2\varphi\,d\varphi.$$

Averaging of the periodic function over the period gives the same result here as averaging (7.30) of the same function in Sec. 7.2 over an infinite interval. Therefore,

$$q_1 = \frac{1}{\pi a}\frac{a}{2}\Big(1-\frac{3}{4}a^2\Big) = \frac{1}{2\pi}\Big(1-\frac{3}{4}a^2\Big).$$

Equation (8.15) takes the form:

$$\frac{d^2 i}{dt^2} - \mu\frac{1}{2\pi\Omega}\Big(1-\frac{3}{4}a^2\Big)\frac{di}{dt} + i = 0.$$

The corresponding characteristic equation is:

$$L(p) = p^2 - \mu\frac{1}{2\pi\Omega}\Big(1-\frac{3}{4}a^2\Big)p + 1 = 0.$$

The condition of existence of the periodic solution according to (8.12) is:

$$X = (1-\omega^2)\Big|_{\omega=\Omega} = 1 - \Omega^2 = 0,$$

$$Y = \mu\frac{\omega}{2\pi\Omega}\Big(1-\frac{3}{4}a^2\Big)\Big|_{\omega=\Omega} = \frac{\mu}{2\pi}\Big(1-\frac{3}{4}a^2\Big).$$

Therefore, $\Omega_0 = 1$, $a_0 = 2/\sqrt{3}$. The stability of this solution is investigated by using (8.13).

$$\Big(\frac{\partial X}{\partial a}\Big)_0\Big(\frac{\partial Y}{\partial\omega}\Big)_0 - \Big(\frac{\partial X}{\partial\omega}\Big)_0\Big(\frac{\partial Y}{\partial a}\Big)_0 = \frac{3\mu a_0}{\pi} > 0.$$

Hence the found periodic motion corresponds to the stable periodic motion similar to Sec. 7.2.

8.2.2. Angular motion of a servomechanism is investigated with taking into account viscous and dry friction on the axis of the shaft and the inertia of the chain of amplification. The equations of this system have the form

$$\begin{gathered} I\frac{d^2\theta}{dt^2} = -R\frac{d\theta}{dt} - F\Big(\frac{d\theta}{dt}\Big) + Cu, \\ \tau\frac{du}{dt} + u = -k\theta, \qquad F\Big(\frac{d\theta}{dt}\Big) = F_0\,\mathrm{sgn}\Big(\frac{d\theta}{dt}\Big), \end{gathered} \tag{8.16}$$

where θ is the rotation angle of the output shaft; u is the voltage at the input of the engine; I is the moment of inertia; R is the coefficient of viscous friction; $F(d\theta/dt)$ is the relay characteristic of dry friction; C is the coefficient of the engine with respect to the control voltage tension u; and k, τ are the coefficients of amplification and the time constant of the amplifier in the feedback chain. Linearization of the system (8.16) should be fulfilled. The linearization coefficient of the relay function F is taken from the table of linearization coefficients for typical nonlinearities [23]. The system (8.16) takes the form:

$$I\frac{d^2\theta}{dt^2} + R\frac{d\theta}{dt} + \frac{4F_0}{2\pi\Omega}\frac{d\theta}{dt} - Cu = 0,$$
$$\tau\frac{du}{dt} + u + k\theta = 0. \tag{8.17}$$

The characteristic polynomial for the system (8.17) has the form:

$$L(p) = I\tau p^3 + \Big(R\tau + \frac{4F_0}{2\pi\Omega}\tau + I\Big)p^2 + \Big(R + \frac{4F_0}{2\pi\Omega}\Big)p + K,$$
$$K = ck.$$

Therefore,

$$X = -\Big(R\tau + \frac{4F_0}{2\pi\Omega}\tau + I\Big)\omega^2 + K,$$
$$Y = -I\tau\omega^3 + \Big(R + \frac{4F_0}{2\pi\Omega}\Big)\omega. \tag{8.18}$$

From the equation $Y\big|_{\omega=\Omega} = 0$ in accordance with (8.18) it is found that

$$\Omega^2 = \frac{R + 4F_0/\pi a\Omega}{I\tau}. \tag{8.19}$$

By excluding Ω^2 from the equation $Y\big|_{\omega=\Omega} = 0$, it is obtained that

$$\Big(R\tau + \frac{4F_0}{2\pi\Omega}\tau + I\Big)\frac{R + 4F_0/\pi a\Omega}{I\tau} + K = 0. \tag{8.20}$$

In Fig. 10 the graph of the dependence of a_0 on K is drawn in accordance with (8.20). In (8.20) the value of Ω is taken to be equal to the root Ω_0 of Eq (8.19). From (8.20) it is readily seen that $K \to \infty$ as $a \to 0$. As $a \to \infty$ the system coincides with the linear system obtained from (8.17) with $F_0 = 0$. The value of coefficient K corresponding to the boundary of the linear system stability is denoted by K_0. Thus $K \to K_0$ as $a \to \infty$.

Let us analyze stability. Taking into account (8.13), (8.18) produces

$$\Big(\frac{\partial X}{\partial a}\Big)_0\Big(\frac{\partial Y}{\partial \omega}\Big)_0 - \Big(\frac{\partial X}{\partial \omega}\Big)_0\Big(\frac{\partial Y}{\partial a}\Big)_0 = -\frac{4F_0}{\pi a_0^2}\Omega_0^2\Big(3I\tau\Omega_0 + \frac{4F_0\tau}{\pi a_0\Omega_0} + 2I\Big) < 0.$$

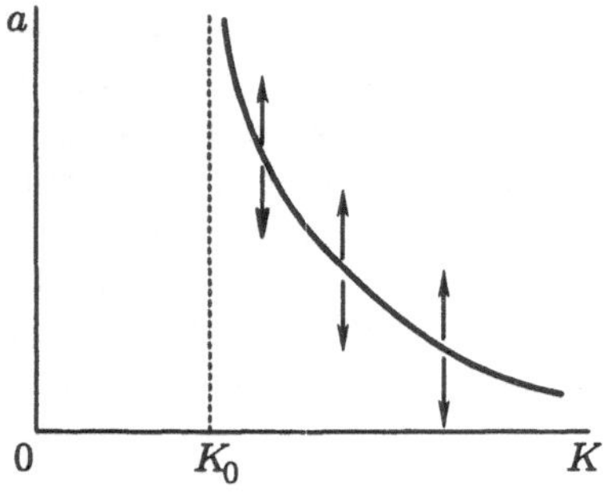

Fig. 10

The periodic solution is unstable; it is the boundary for the initial conditions. In Fig. 10 the direction of the change of the variable a is shown by arrows.

Thus the method of harmonic linearization makes it possible to determine the amplitude of the periodic solution of the system and behavior of small deviations from the amplitude; that is, it determines the behavior of the system with respect to the slow variable a. In this sense the method of harmonic linearization is one of the varieties of the method of motion decomposition.

9 Method of averaging in systems with several fast phases

9.1 Averaged equations of the first approximation.

Systems with several fast phases resemble:

$$\begin{aligned} \frac{dy}{dt} &= \varepsilon Y(y, z, \varepsilon), & y(0) &= y_0, \\ \frac{dz}{dt} &= \omega(y) + \varepsilon Z(y, z, \varepsilon), & z(0) &= z_0. \end{aligned} \tag{9.1}$$

In contrast to (6.10), here y is a vector of arbitrary dimension m and z is a vector of dimension n. Similar to (6.10), assume that functions are 2π-periodic with respect to all fast variables $z_1, \dots, z_n$. If a system has different $T_1, \dots, T_n$ periods for different variables $z_1, \dots, z_n$, then we can represent it in the 2π-periodic form in all variables, rescaling $z_k = (T_k/2\pi)w_k$.

We can compute successive approximations to the solution of (9.1) it the same way as in Sec. 6 [13]. But formulas with z, $\overline{z}$, $\widetilde{z}$ in this case are very complicated.

Consider the averaged equation of the first approximation (see (6.27)).

$$\frac{d\overline{y}}{dt} = \varepsilon \overline{Y}^{(0)}(\overline{y}), \qquad \overline{y}(0) = y_0. \tag{9.2}$$

Function $\overline{Y}^{(0)}$ is obtained from the same condition as in Sec. 6: norms of the functions $||\widetilde{y}||$, $||\widetilde{z}||$ should be bounded. The following formulae give the same result in a much more complicated way.

$$\overline{Y}^{(0)} = \frac{1}{(2\pi)^n} \int_0^{2\pi} \cdots \int [Y^{(0)}(\overline{y}, z_1, \ldots, z_n)] \, dz_1 \ldots dz_n \,. \tag{9.3}$$

Expression (9.3) is similar to (6.26) but the mean value is computed in (9.3) for all fast phases.

The complex Fourier-series expansion for the 2π-periodic in $z_1, \ldots, z_n$ function $Y^{(0)}$ is as follows.

$$Y^{(0)}(y, z_1, \ldots, z_n) = \sum_{||l|| \geq 0} Y_l^{(0)}(y) e^{i(l,z)}. \tag{9.4}$$

Here $Y_l^{(0)}$ are the coefficients of the expansion and $(l, z) = l_1 z_1 + \ldots + l_n z_n$ is the scalar product of the vector $z(z_1, \ldots, z_n)$ of fast phases to the vector $l(l_1, \ldots, l_n)$ with arbitrary integer components such that $||l|| \geq 0$.

Let us substitute (9.4) into (9.3). After averaging all harmonics vanish and $\overline{Y}^{(0)} = \overline{Y}_0^{(0)}$, where $\overline{Y}_0^{(0)}$ is the zeroth term of the Fourier-series expansion of (9.4).

In the papers by Grebennikov [20] and Volosov and Morgunov [21], the authors show that Eqs. (9.2) and (9.3) define the slow variable y with an error of order ε at the time interval $t \sim 1/\varepsilon$.

When z is a vector, the construction of averaged equations demands hard work for computing multiple integrals (9.3). The operation of averaging along trajectories of a generating system make calculations easy; only single integration with respect to t is necessary. In this case calculations are the same as in Sec. 6.4; the only difference is that z, z_0, ω in (6.38) and (6.40) are now the vectors.

Now consider a formidable danger for a scientist while analyzing averaged equations with multiple fast phases.

9.2 Resonances in multifrequency systems

Let us compute the rightmost part of averaged equation (9.2), using the method of averaging along the trajectories of a generating system. Substituting (9.4) into (6.40), where $z = z_0 + \omega t$ is the vector of the solution of the generating system, we get:

$$\frac{dz}{dt} = \omega(y), \quad z(0) = z_0, \quad y = \text{const}$$

for (9.1). Then

$$\overline{Y}^{(0)} = \lim_{L\to\infty} \frac{1}{L} \int_0^L \Big[\sum_{\|l\|\geq 0} Y_l^{(0)}(y) e^{i(l,z_0+\omega t)} \Big]\, dt. \tag{9.5}$$

The general term of the series in (9.5) contains the multiplier:

$$e^{i(l,z_0+\omega t)} = e^{i(l,z_0)} e^{i(l_1\omega_1+\ldots+l_n\omega_n)t}. \tag{9.6}$$

Integration of (9.6) produces:

$$\frac{-ie^{i(l,z_0)}e^{i(l,\omega)t}}{(l_1\omega_1+\ldots+l_n\omega_n)}.$$

It is readily seen that, under the condition

$$(l,\omega) = l_1\omega_1 + \ldots + l_n\omega_n = 0 \tag{9.7}$$

the operation (9.5) has no sense. One says that there is resonance in the system. To check whether a system can have a resonance, the values $\omega_1, \ldots, \omega_n$ in (9.7) are taken from initial equations (9.1), and $l_1, \ldots, l_n$ are sorted on the set of integers. Dependence of the result of the averaging (9.5) on initial data by $z_1, \ldots, z_n$ also may be regarded as a formal criterion of resonance appearance. If (9.7) is not satisfied, then all the harmonic components depending on initial data would vanish in the course of averaging.

It should be noted that in the case of resonance both operation (9.5) and averaging by phases (9.3) make no sense. In this situation expressions of the type (9.7) also appear in denominators.

9.3 Averaging algorithm in the case of resonance

Suppose that for a set of frequencies $\omega_1, \ldots, \omega_n$ of the system (9.1) there exists a set of numbers $l_1^0, \ldots, l_n^0$, such that the condition of resonance (9.7) is satisfied. For this fixed set $l_1^0, \ldots, l_n^0$ the values $\omega_1, \ldots, \omega_n$, near the values from (9.7) are considered, such that the condition (9.7) is satisfied approximately:

$$l_1^0\omega_1 + \ldots + l_n^0\omega_n = \varepsilon\Delta(y_1, \ldots, y_n). \tag{9.8}$$

Here $\varepsilon\Delta$ is called a frequency deviation, factor $\varepsilon \ll 1$ indicates a small value of the deviation, and $\Delta(y_1, \ldots, y_n) \sim 1$.

It should be noted that in general the values $\omega_1, \ldots, \omega_n$ depend on slow variables $y_1, \ldots, y_m$. Therefore, the value of the deviation may be changed when the variables $y_1, \ldots, y_m$ are changed. If the value $\varepsilon\Delta$ increases then the system goes out of resonance. Due to changes of $y_1, \ldots, y_m$ the system can reach a new resonance, corresponding to another set of numbers $l_1, \ldots, l_n$. Investigation of these problems is very difficult and far from completeness [20]. These problems are not discussed in our book.

Expression (9.8) should be compared with a linear combination of variables:

$$l_1^0 z_1 + \ldots + l_n^0 z_n = \delta. \tag{9.9}$$

A value δ is called a phase deviation. Comparison of (9.8) and (9.9) shows that the rate of change of the variable δ is the value of the order ε; that is, motion of the system in the proximity of resonance is described by this slow variable.

In (9.1) let us change a set of initial phase variables $y_1, \ldots, y_m; z_1, \ldots, z_n$ to a new set $y_1, \ldots, y_m, \delta; z_1, \ldots, z_{n-1}$ excluding with the aid of (9.9) one of the variables z_j, say, z_n:

$$\begin{aligned}
\frac{dy_1}{dt} &= \varepsilon Y_1\Big(y_1, \ldots, y_m; z_1, \ldots, z_{n-1}, \frac{\delta - l_1^0 z_1 - \ldots - l_{n-1}^0 z_{n-1}}{l_n^0}\Big), \\
&\ldots\ldots \\
\frac{dy_m}{dt} &= \varepsilon Y_m(\ldots), \\
\frac{dz_1}{dt} &= \omega_1 + \varepsilon Z_1\Big(y_1, \ldots, y_m; z_1, \ldots, z_{n-1}, \frac{\delta - l_1^0 z_1 - \ldots - l_{n-1}^0 z_{n-1}}{l_n^0}\Big), \\
&\ldots\ldots \\
\frac{dz_{n-1}}{dt} &= \omega_{n-1} + \varepsilon Z_{n-1}(\ldots).
\end{aligned} \tag{9.10}$$

The differential equation for δ is obtained by differentiating (9.9) and substituting the derivatives (dz_j/dt) by their expressions from (9.1)

$$\frac{d\delta}{dt} = l_1^0 \frac{dz_1}{dt} + \ldots + l_n^0 \frac{dz_n}{dt} = (l_1^0 \omega_1 + \ldots + l_n^0 \omega_n) + \varepsilon(l_1^0 Z_1 + \ldots + l_n^0 Z_n). \tag{9.11}$$

Taking into account (9.8), we transform Eq. (9.11) into the form:

$$\frac{d\delta}{dt} = \varepsilon\Delta + \varepsilon(l_1^0 Z_1 + \ldots + l_n^0 Z_n). \tag{9.12}$$

In the right-hand side of (9.12) the variable z_n should be excluded, as in (9.10).

Equations (9.10) and (9.12) form a closed system of equations with respect to slow variables $y_1, \ldots, y_m, \delta$ and fast variables $z_1, \ldots, z_{n-1}$. By analogy with (9.7), resonance assumptions in this system are:

$$\widetilde{l}_1 \omega_1 + \ldots + \widetilde{l}_{n-1} \omega_{n-1} = 0. \tag{9.13}$$

Generally speaking, the set of resonance numbers $l_1^0, \ldots, l_{n-1}^0$ of the initial system (9.1) does not coincide with the numbers $\widetilde{l}_1, \ldots, \widetilde{l}_{n-1}$ from (9.13). For this reason the system (9.10) and (9.12) for the initial set $l_1^0, \ldots, l_{n-1}^0$ has no resonance and may be investigated according to the nonresonance averaging scheme from Sec. 8.1.

9.4 Pendulum resonance oscillations with horizontal and vertical vibration of the point of suspension

9.4.1 Horizontal vibration.

Now it is assumed that a point of suspension O of a physical pendulum moves with acceleration

$$W_{0x} = W_0 \sin \omega_1 T, \tag{9.14}$$

where W_0 and ω_1 are the amplitude and frequency of the vibration. The equation of pendulum motion has the form:

$$I\ddot{\Psi} = -R\dot{\Psi} - lM(g \sin \Psi + W_{0x} \cos \Psi), \tag{9.15}$$

where Ψ is an angle of the deviation from the vertical, M, I, l are mass, moment of inertia, and distance from the center of the mass to the point of suspension, and R is the coefficient of friction.

Substitution in (9.15) of $\Omega = d\Psi/dT$ produces a linear system of first-order equations. It should be normalized. A class of small oscillations of the pendulum with small friction at the resonance frequency of the disturbance is considered. To estimate characterizing values, the linear equation corresponding to (9.15) is examined:

$$\begin{gathered} I\ddot{\Psi} + R\dot{\Psi} + K\Psi = -M_0 \sin \omega_1 T, \\ K = lMg, \qquad M_0 = lMW_0 \,. \end{gathered} \tag{9.16}$$

When the friction is small a characteristic frequency ω_0 for (9.16) is estimated by the formula $\omega_0^2 = K/I$. When the resonance $\omega_1 = \omega_0$ takes place, the moment of inertial force and the restoring pendular moment delete each other in (9.16), and the characterizing value of the angular velocity is determined by the estimate:

$$R\Omega_* = M_0. \tag{9.17}$$

As previously, for harmonic oscillations with frequency near to ω_0, characteristic time is $T_* = T_0$, and characterizing amplitudes of angle and of angular velocity are connected by the relation: $\Omega_* = \Psi_* \omega_0$. Let the characterizing angle be small: $\Psi_* \ll 1$. After expanding the right-hand parts of (9.15) into a series with respect to Ψ, Eqs. (9.15) and (9.14) take the form:

$$\begin{aligned} \frac{d\omega}{dt} &= -2\zeta\omega - \left(\psi - \frac{1}{6}\Psi_*^2\psi^3 + \ldots\right) - 2\zeta\left(1 - \frac{1}{2}\Psi_*^2\psi^2\right) \sin \nu t, \\ \frac{d\psi}{dt} &= \omega \,. \end{aligned} \tag{9.18}$$

Here 2ζ is the dimensionless coefficient of damping, and $\nu = \omega_1 T_0$ is the dimensionless frequency of perturbation. Equations (9.18) have two small parameters 2ζ and Ψ_*. From (9.17) it follows that they are connected with the relation:

$$2\zeta\Psi_* = \Psi_0, \tag{9.19}$$

where $\Psi_0 = M_0/K$ is the static deviation of the system (9.15). In the following 2ζ and Ψ_0 are assumed to be independent dimensionless parameters, which characterize the damping of the system and intensity of perturbation. Depending on relative quantities of these parameters, the right-hand sides of (9.19) may depend on small parameters in a different way, and approximate solutions are also different. The simplest case from a computational point of view is considered when the parameters 2ζ and Ψ_*^2 in (9.18) are of the same order:

$$\frac{\Psi_*^2}{6} = e2\zeta, \qquad e \sim 1\,. \tag{9.20}$$

Let us exclude in (9.20) the value Ψ_* using (9.19). Hence evaluation $2\zeta^3 \sim \Psi_0^2$ is obtained, which determines the class of motion, where considered asymptotic is acceptable.

Let us account for members up to the first order in (9.18). Then

$$\begin{aligned} \frac{d\omega}{dt} &= -\psi + \varepsilon f, \\ \frac{d\psi}{dt} &= \omega, \end{aligned} \tag{9.21}$$

where

$$\varepsilon = 2\zeta \ll 1, \quad f = -\omega + e\psi^3 - \sin\alpha, \quad \alpha = \nu t. \tag{9.22}$$

As previously, fast and slow variables in (9.21) are separated by substitution

$$\psi = a\sin\varphi, \qquad \omega = a\cos\varphi\,. \tag{9.23}$$

Then (9.21) and (9.22) take the form

$$\begin{aligned} \frac{da}{dt} &= \varepsilon(-a\cos\varphi + ea^3\sin^3\varphi - \sin\alpha)\cos\varphi, \\ \frac{d\varphi}{dt} &= 1 - \frac{\varepsilon}{a}(-a\cos\varphi + ea^3\sin^3\varphi - \sin\alpha)\sin\varphi, \\ \frac{d\alpha}{dt} &= \nu\,. \end{aligned} \tag{9.24}$$

System (9.24) has two frequencies $\omega_1 = 1$, $\omega_2 = \nu$. The resonance condition for this system is:

$$l_1\omega_1 + l_2\omega_2 = l_1\cdot 1 + l_2\nu = 0\,. \tag{9.25}$$

The case of the so-called main resonance $\nu = 1$ is considered. Then Eq. (9.25) holds for $l_1^0 = 1$, $l_2^0 = -1$. Let us investigate a motion of the system (9.24) in the proximity of this resonance following the methods of Sec. 9.3. Let us introduce according to (9.8) the frequency deviation

$$\varepsilon\Delta = 1 - \nu, \tag{9.26}$$

and in accordance with (9.9), the phase deviation

$$\delta = \varphi - \alpha. \tag{9.27}$$

The variable φ is excluded from (9.24) using (9.27). Taking into account (9.26) produces

$$\begin{aligned}
\frac{da}{dt} &= \varepsilon[-a\cos^2(\alpha+\delta) + ea^3\sin(\alpha+\delta)\cos(\alpha+\delta) - \cos(\alpha+\delta)\sin\alpha],\\
\frac{d\delta}{dt} &= \varepsilon\Delta + \varepsilon[\frac{1}{2}\sin 2(\alpha+\delta) - ea^2\sin^4(\alpha+\delta) - \frac{1}{a}\sin(\alpha+\delta)\sin\alpha],\\
\frac{d\alpha}{dt} &= \nu.
\end{aligned} \tag{9.28}$$

In the system (9.28) there is one fast variable α; therefore, resonance does not exist.

Let us average the right-hand sides of (9.28) over α, assuming the slow variables a, δ to be parameters. Averaged equations of the first approximation are:

$$\begin{aligned}
\frac{d\bar{a}}{dt} &= \varepsilon\Big(-\frac{\bar{a}}{2} + \frac{1}{2}\sin\bar{\delta}\Big),\\
\frac{d\bar{\delta}}{dt} &= \varepsilon\Big(\Delta + \frac{1}{2\bar{a}}\cos\bar{\delta} - \frac{3}{8}e\bar{a}^2\Big).
\end{aligned} \tag{9.29}$$

Now stationary solutions $\bar{a}, \bar{\delta} = \text{const}$ of the system (9.29) should be considered. They are defined by the equations:

$$\sin\bar{\delta} = \bar{a}, \qquad \cos\bar{\delta} = -2\bar{a}\,(\Delta - \frac{3}{8}e\bar{a}^2). \tag{9.30}$$

By excluding in (9.30) the variable $\bar{\delta}$, an equation of the resonance curve $\bar{a}\,(\Delta)$ is obtained:

$$F(\bar{a}^2, \Delta) \equiv \bar{a}^2 + 4\bar{a}^2(\Delta - \frac{3}{8}e\bar{a}^2) - 1 = 0. \tag{9.31}$$

Let us define frequency providing an extremum to the function $\bar{a}^2(\Delta)$. Taking into account the equation

$$\frac{d\bar{a}^2}{d\Delta} = -\frac{\partial F}{\partial\Delta}\Big/\frac{\partial F}{\partial\bar{a}^2} = 0,$$

conclude that the extremum is defined by the equation:

$$\frac{\partial F}{\partial \Delta} = 8\,\overline{a}^{\,2}(\Delta - \frac{3}{8}e\,\overline{a}^{\,2}) = 0;$$

hence,

$$\Delta_{\max} = \frac{3}{8}e\,\overline{a}^{\,2}. \tag{9.32}$$

From (9.31) and (9.32) an extremal value of the amplitude $\overline{a}^{\,2}_{\max} = 1$, equal to the height of resonance pick is obtained for a linear system.

From (9.26) and (9.32) one can see that the resonance pick of the system (9.28) corresponds to the frequency

$$\nu = 1 - \varepsilon\Delta = 1 - \frac{3}{8}\varepsilon e\,\overline{a}^{\,2}. \tag{9.33}$$

Thus, the resonance pick of the system with the "soft" characteristic of the nonlinear restoring force (see. Sec. 5.2.) is displaced on the left of resonance frequency $\nu = 1$ of the corresponding linear system.

In (9.33) variables having dimension will be used. Taking into account (9.18), (9.20), and (9.22), the following equation is obtained.

$$\omega_1 = \omega_0\Big(1 - \frac{1}{16}\Psi_*^2\Big). \tag{9.34}$$

Comparison of (9.34) with (5.15) demonstrates that a shift of frequency of the resonance picks depending on amplitude Ψ_* of forced oscillations coincides exactly with a shift of frequency of free oscillations. Thereby, picks of resonance curves depending on changes of parameter Ψ_* lie on the parabola given by Fig. 4. According to Sec. 5.2 these picks lie on the so-called skeleton curve of the family of resonance characteristics by parameter Ψ_*. From (9.17) one can see that the quantity Ψ_* is proportional to the amplitude of the disturbance M_0. Therefore, the skeleton curve by the parameter M_0 is also parabolic.

A case of a "hard" characteristic of restoring force can be obtained from (9.21) and (9.22) with $e < 0$. From (9.33) one can see that in this case the resonance pick is displaced to the right.

9.4.2. Vertical vibration.

Let vertical vibration of the suspension point be $W_{0Y} = W_0 \sin\omega_1 T$. For simplicity let us consider a motion without friction. Equation of the motion for a pendulum has the form

$$I\ddot{\Psi} = -lM(g + W_0\sin\omega_1 T)\sin\Psi\,. \tag{9.35}$$

Normalizing the system, suppose the motion to be near the harmonic. Then, as previously, $T_* = T_0 = 1/\omega_0$, $\Omega_* = \Psi_*/T_0$, where $\omega_0^2 = Mgl/I$.

Normalization brings (9.35) into the form

$$\begin{aligned} \Psi_* \frac{d\omega}{dt} &= -\left(1 + \frac{W_0}{g} \sin \nu t\right) \sin \Psi_* \psi, \\ \frac{d\psi}{dt} &= \omega, \end{aligned} \tag{9.36}$$

where W_0/g and $\nu = \omega_1 T_0$, the depth of modulation of the coefficient from (9.36) and the dimensionless frequency of disturbance.

In (9.36) suppose W_0/g and Ψ_* to be small. Without consideration of possible relations between these parameters, members of first order by them are accounted for in (9.36). Then in (9.36) parameter Ψ_* disappears, and equations take a form:

$$\begin{aligned} &\frac{d\omega}{dt} = -\psi + \varepsilon f, \\ &\frac{d\psi}{dt} = \omega, \qquad \frac{d\alpha}{dt} = \nu, \end{aligned} \tag{9.37}$$

where

$$\varepsilon = W_0/g \ll 1, \qquad f = -\psi \sin \alpha. \tag{9.38}$$

As in Sec. 8.4.1, substitution

$$\psi = a \sin \varphi, \qquad \omega = a \cos \varphi$$

transforms (9.37) and (9.38) into the following system

$$\begin{aligned} &\frac{da}{dt} = \varepsilon(-a \cos \varphi \sin \alpha) \cos \varphi, \\ &\frac{d\varphi}{dt} = 1 - \frac{\varepsilon}{a}(-a \sin \varphi \sin \alpha) \sin \varphi, \qquad \frac{d\alpha}{dt} = \nu. \end{aligned} \tag{9.39}$$

System (9.39) has two frequencies: $\omega_1 = 1, \omega_2 = \nu$. The resonance assumption for it is:

$$l_1 \omega_1 + l_2 \omega_2 = l_1 1 + l_2 \nu = 0. \tag{9.40}$$

Let us consider a case of multiple resonance $\nu = 2$. Then the condition (9.40) is true with $l_1^0 = 2$, $l_2^0 = -1$. Let us introduce a frequency deviation

$$\varepsilon \Delta = 2 - \nu \tag{9.41}$$

and a phase deviation

$$\delta = 2\varphi - \alpha. \tag{9.42}$$

With the aid of (9.41) and (9.42) rewrite (9.38) in the form

$$\begin{aligned} \frac{da}{dt} &= -\varepsilon a \frac{1}{2} \sin(\alpha + \delta) \sin \alpha, \\ \frac{d\delta}{dt} &= \varepsilon \Delta + \varepsilon \frac{1}{2} [1 - \cos(\alpha + \delta)] \sin \alpha, \\ \frac{d\alpha}{dt} &= \nu. \end{aligned} \tag{9.43}$$

System (9.43) is already nonresonant.

For this system averaged by α, first approximation equations have the following form

$$\begin{aligned} \frac{d\bar{a}}{dt} &= -\varepsilon \bar{a} \frac{1}{4} \cos \bar{\delta}, \\ \frac{d\bar{\delta}}{dt} &= \varepsilon \left(\Delta + \frac{1}{4} \sin \bar{\delta}\right). \end{aligned} \tag{9.44}$$

From (9.44) equations for stationary solutions $\bar{a}, \bar{\delta} = \text{const}$ are obtained

$$\bar{a} \cos \bar{\delta} = 0, \qquad \Delta + \frac{1}{4} \sin \bar{\delta} = 0. \tag{9.45}$$

Hence $\bar{\delta} = \pm\pi/2$, $\Delta = \pm 1/4$.

Let us write the second equation from (9.45) in initial parameters. From (9.38) and (9.41) we get

$$(2 - \nu) + \frac{1}{4} \frac{W_0}{g} \sin \bar{\delta} = 0. \tag{9.46}$$

Equation (9.46) with $\bar{\delta} = \pm\pi/2$ in the plane $(\nu, W_0/g)$ gives a linear approximation of two branches of the Ince–Strutt diagram for the periodic solution of the Mathieu equation (9.37).

9.5 Resonant oscillations with friction of different kinds

9.5.1 Linear friction

A mechanical system with one degree of freedom is considered, which moves under the action of a linear viscoelastic force and harmonic perturbation with an amplitude F_0 and frequency ω_1. An equation of motion is obtained from (7.1) by adding the perturbation:

$$M\ddot{x} + R\dot{x} + Kx = F_0 \sin \omega_1 T.$$

In this case conclusions about resonant behavior can be obtained from the results of Sec. 9.4.1. The formula (9.19) after obvious denotations take the form:

$$2\zeta X_* = F_0/K. \tag{9.47}$$

In (9.21) the multiplier e at the cubic term of positional force should be set equal to zero. Then from (9.31) and (9.32) the shift of the resonant peak with respect to the frequency $\Delta_{\max} = 0$ and the amplitudes of the resonant oscillations $\overline{a}_{\max} = 1$ are obtained. After transition to dimensional quantities from (9.33) the resonant frequency of perturbation is obtained: $\omega_1 = \omega_0$, and from (9.47) the amplitude $A_{\max} = X_* a_{\max} = F_0/2\zeta K$ is found.

9.5.2 Quadratic friction.

An equation of motion is obtained from (7.12) by adding an harmonic perturbation:

$$M\ddot{x} = -Kx - \varrho\frac{\dot{x}^2}{2}C_x S\,\mathrm{sgn}\,\dot{x} + F_0 \sin\omega_1 T. \tag{9.48}$$

The perturbation close to the resonant one is considered: $(\omega_1 \approx \omega_0)$. Friction is assumed to be small. Then by analogy with (7.1) and (7.12), while normalizing the system (9.48) it can be assumed that $V_* = X_*\omega_0$, $T_* = 1/\omega_0$. For estimating the value of X_* it is taken into account that in the proximity of the resonance, elastic and inertial forces are balanced. The estimating equation for remaining forces, that is, for the forces of friction and perturbation, has the form:

$$\frac{1}{2}\varrho C_x S(X_*\omega_0)^2 = F_0.$$

Therefore,

$$\lambda X_*^2 = F_0; \qquad \lambda = \varrho C_x S\omega^2/2. \tag{9.49}$$

After normalization the system (9.48) is reduced to the form:

$$\begin{aligned} &\frac{dv}{dt} = -x + \varepsilon f, \qquad \frac{dx}{dt} = v, \qquad \frac{d\alpha}{dt} = \nu, \\ &f = -v|v| + \sin\alpha, \qquad \varepsilon = \frac{\varrho C_x S X_*}{M}, \qquad \nu = \frac{\omega_1}{\omega_0}. \end{aligned} \tag{9.50}$$

The class of motion is considered for which the condition $\varepsilon \ll 1$ is satisfied. For (9.50) the averaged equation of the first approximation should be constructed. The calculations by analogy with Sec. 9.3 and 9.4 produce:

$$\begin{aligned} \frac{d\overline{a}}{dt} &= -\varepsilon\Big[\frac{4}{3\pi}\overline{a}^2 + \frac{1}{2}\sin\overline{\delta}\Big], \\ \frac{d\overline{\delta}}{dt} &= \varepsilon\Big[\Delta - \frac{1}{2\overline{a}}\cos\overline{\delta}\Big]. \end{aligned} \tag{9.51}$$

Stationary solutions of the system (9.51) satisfy the following equation obtained by setting $\bar{a}$, $\bar{\delta}$ = const in (9.51) and excluding the variable $\bar{\delta}$:

$$F = -\frac{1}{4} + \left(\frac{4}{3\pi}\bar{a}^2\right)^2 + \left(\Delta^2\bar{a}^2\right) = 0. \tag{9.52}$$

The frequency determined by the equation $\partial F/\partial\Delta = 0$ corresponds to the resonant peak. Therefore, $\Delta_{\text{max}} = 0$. Then the resonant amplitude is obtained from (9.52) by the formula:

$$\bar{a}^2_{\text{max}} = \frac{3}{8}\pi. \tag{9.53}$$

After transition to dimensional quantities from (9.49) and (9.53) the amplitude of resonant oscillations is obtained:

$$A_{\text{max}} = X_* \, \bar{a}_{\text{max}} = \sqrt{\frac{3\pi}{8\lambda}F_0}.$$

9.5.3 Coulomb friction.

Now the equations of motion additionally contain harmonic perturbation as compared with the equations in Sec. 7.1.3:

$$M\ddot{x} = -Kx - R_0 \operatorname{sgn}\dot{x} + F_0 \sin\omega_1 T. \tag{9.54}$$

It is required that the following condition be satisfied:

$$F_0 > R_0\,. \tag{9.55}$$

In the opposite case the system is in the stagnation zone $|X| \le X_R = R_0/K$.

After normalizing the system (9.54) by analogy with Sec. 9.5.1 and 9.5.2, it can be written in the form:

$$\begin{aligned}
\frac{dv}{dt} &= -x + \varepsilon f, & f &= -\frac{|v|}{v} + \lambda\sin\alpha,\\
\frac{dx}{dt} &= v, & \varepsilon &= \frac{R_0}{KX_*} = \frac{X_R}{X_*},\\
\frac{d\alpha}{dt} &= \nu, & \lambda &= \frac{X_{\text{st}}}{X_R},
\end{aligned} \tag{9.56}$$

where $X_{\text{st}} = F_0/K$ is the value of the static deviation and $X_R = R_0/K$ is the size of the stagnation zone.

The class of motion is considered for which the condition $\varepsilon \ll 1$, $\lambda \sim 1$ is satisfied. For (9.56) the averaged equations of the first approximation take the form:

$$\begin{aligned}
\frac{d\bar{a}}{dt} &= \varepsilon\Big[-\frac{2}{\pi} - \frac{\lambda}{2}\sin\bar{\delta}\Big],\\
\frac{d\bar{\delta}}{dt} &= \varepsilon\Big[\Delta - \frac{\lambda}{2\bar{a}}\cos\bar{\delta}\Big].
\end{aligned} \tag{9.57}$$

Stationary solutions of the system (9.57) satisfy the equations similar to (9.52)

$$F = -\frac{\lambda^2}{4} + \frac{4}{\pi^2} + \Delta^2 \overline{a}^{\,2}.$$

Therefore,

$$\overline{a}^{\,2} = \frac{\lambda^2/4 - 4/\pi^2}{\Delta^2}. \tag{9.58}$$

From (9.58) it is readily seen that $\overline{a}^{\,2} \to \infty$ as $\Delta^2 \to 0$. Hence, for the resonant frequency of perturbation Coulomb friction does not bound the amplitude of forced oscillations.

The right-hand side of formula (9.58) should be positive. Therefore, the restriction $\lambda = F_0/R_0 \geq 4/\pi = 1.27$ is obtained. This inequality coincides with the condition (9.55).

10 Averaging in systems without explicit periodicities

10.1 Volosov averaging scheme

The system of general form containing both fast and slow variables is considered:

$$\begin{aligned} \frac{dy}{dt} &= \varepsilon Y(y, z, t, \varepsilon), \qquad & y(0) &= y_0, \\ \frac{dz}{dt} &= Z(y, z, t, \varepsilon), \qquad & z(0) &= z_0, \end{aligned} \tag{10.1}$$

where $\varepsilon \ll 1$ is a small parameter, y, z are vectors of slow and fast variables of arbitrary dimensions, and t is fast dimensionless time. However, the last may be included in the list of fast variables as in (6.9). It is assumed that Y, Z are analytical functions of their arguments; therefore, expansions of the right-hand sides of (10.1) in series with respect to ε have the form:

$$\begin{aligned} \frac{dy}{dt} &= \varepsilon[Y^{(0)}(y, z, t) + \varepsilon Y^{(1)}(y, z, t) + \ldots], \\ \frac{dz}{dt} &= Z^{(0)}(y, z, t) + \varepsilon Z^{(1)}(y, z, t) + \ldots . \end{aligned} \tag{10.2}$$

Equations (6.9), (6.10), and (6.1) analyzed earlier with the aid of averaging methods have some restrictive features. The main ones are:

(1) the zero member of the expansion in the equations in the fast variables $Z^{(0)} \equiv \omega$ was supposed to depend only on slow variables $y_1, \ldots, y_m$;

(2) the functions $Y^{(0)}, Y^{(1)}, \ldots; Z^{(0)}, Z^{(1)}, \ldots$ were supposed to be 2π-periodic in fast variables.

These restrictions may be avoided. For the system of general form (10.1) the scheme of averaging is proposed that uses the scheme of averaging along the trajectories of generating systems [21]. This scheme allows us to implement an approximated decomposition of motions.

The generating system of equations for (10.1), which is obtained from it with $\varepsilon = 0$, is:

$$\frac{dz}{dt} = Z^{(0)}(y, z, t), \qquad z(0) = z_0, \qquad y = \text{const}. \tag{10.3}$$

Solution of the system (10.3) is:

$$z = z(y, t, z_0). \tag{10.4}$$

By substituting $Y^{(0)}$ from (10.2) into (10.4), by analogy with (6.40), the average along trajectories of (10.4) is obtained:

$$\overline{Y}^{(0)}(y) = \lim_{L\to\infty} \frac{1}{L} \int_0^L Y^{(0)}(y, z(y, t, z_0), t)\, dt. \tag{10.5}$$

The averaged equation

$$\frac{dy}{dt} = \varepsilon \overline{Y}^{(0)}(y), \qquad y(0) = y_0 \tag{10.6}$$

defines [21] the asymptotic approximation of the solution for y of the system (10.1) in the time interval $t \sim 1/\varepsilon$. In Volosov and Morgunov [21] the restrictions with which the equations (10.1) should comply are formulated.

It should be especially noted that the result of averaging (10.5) should not depend on initial values z_0 of fast variables. As seen from Sec. 9.3, this restriction excludes implicitly the possibility of resonance appearance.

10.2 Separation of characteristic motions of an oscillator with high friction

The equation (7.1) of an oscillating unit with linear friction is considered:

$$M\ddot{X} = -R\dot{X} - KX, \qquad X(0) = X_0, \quad \dot{X}(0) = \dot{X}_0. \tag{10.7}$$

Here, unlike in Sec. 7.1.1, the coefficient of friction R is "large." As it is known, under this condition motion becomes aperiodic, with very differentiating time characteristics of motion. An analogous system was considered in Sec. 2.1.2.

Evaluation of characteristic time T_1 for fast changes of velocity caused by large friction is obtained from the equation $M\ddot{X} + R\dot{X} = 0$, from which $T_1 = M/R$. Slow quasistatic motions along a coordinate are estimated by the equation of balance of forces $R\dot{X} + KX = 0$. Therefore, the estimate for characteristic time of this motion is: $T_2 = R/K$, and the estimate for its characteristic velocity is: $\dot{X}_* = X_*/T_2$. Only the values of parameters are considered for which $T_1 \ll T_2$.

Dividing Eq. (10.7) by K leads to

$$T_1 T_2 \ddot{X} + T_2 \dot{X} + X = 0. \tag{10.8}$$

Let us introduce a variable $V{=}dX/dT$ and perform normalization: $t{=}T/T_*$, $x{=}X/X_*$, $v{=}V/V_*$. It is assumed that $\dot{X}_0(0) = V_0$ has the order of maximal velocity $\dot{X}_* = X_0/T_2$ of quasi-static motion. Let $X_* = X_0$, $V_* = X_*/T_2$. The motion in fast time is considered, as everywhere in chapter III; therefore, it is taken that $T_* = T_1$. Then (10.8) can be written in the form:

$$\begin{aligned} &\frac{dx}{dt} = \mu v, \qquad x(0) = 1, \\ &\frac{dv}{dt} = -v - x, \qquad v(0) = v_0 = \frac{\dot{X}_0}{X_0} T_1, \qquad \mu = \frac{T_1}{T_2} \ll 1 . \end{aligned} \tag{10.9}$$

A system averaged according to Volosov for (10.9) should be constructed. The generating system has the form:

$$\frac{dv}{dt} = -v - x, \qquad v(0) = v_0, \qquad x = \text{const}. \tag{10.10}$$

Its solution is:

$$v(t) = -x + (x + v_0)e^{-t}. \tag{10.11}$$

After substitution of (10.11) into the right-hand side of the first equation of (10.9) the following integral should be calculated

$$\overline{Y}^{(0)}(x) = \lim_{L\to\infty} \frac{1}{L} \int_0^L [-x + (x + v_0)e^{-t}]\, dt = -x. \tag{10.12}$$

The result of averaging (10.12) does not depend on v_0, as is required in accordance with Volosov and Morgunov [21]. An equation of the type of (10.6) averaged according to Volosov has the form:

$$\frac{dx}{dt} = -\mu x, \qquad x(0) = x_0. \tag{10.13}$$

This equation describes the process of slow quasistatic changes of coordinate x. The value of phase variable v on the trajectories of these slow motions can be defined by substitution of (10.13) into the first equation of (10.9).

$$v = -x. \tag{10.14}$$

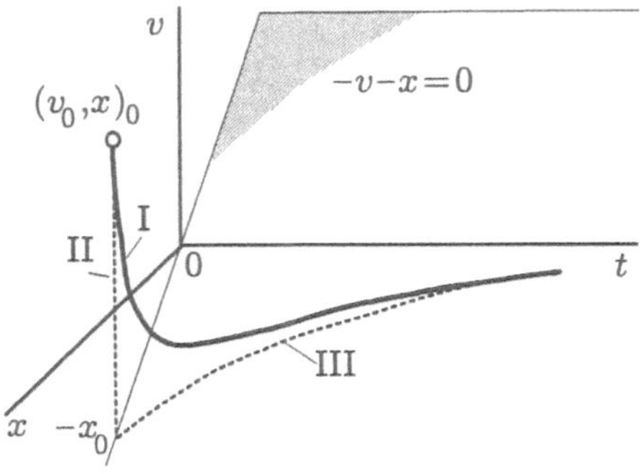

Fig. 11

Let us consider the meaning of a solution of the generating system (10.10) in terms of mechanics. From (10.11) it is seen that Eq. (10.10) describes a process of fast changes of velocity v from its initial value v_0 to the value $v = -x_0$, which corresponds, in accordance with (10.14), to slow motion at an initial moment of time.

In Fig. 11 the type of motion of the system in the space of variables v, x, t is presented. The trajectory of the exact solution of the initial system (10.9) is shown by solid line I; by dotted line II, of the system (10.10); and by dotted line III, of the system (10.13) and (10.14). On the figure the plane $-v - x = 0$ is marked, to which the fast components of the motion tend, according to (10.11). According to (10.14), slow components of the motion are placed in the same plane.

Figure 11 shows that the solution of the system (10.9) may be approximately presented as a result of concatenation of the solutions of simplified systems (10.10), (10.13) and (10.14).

It is interesting to estimate time t_1, after which it is possible to suppose that the system moving along fast trajectory II on Fig. 11 reaches slow trajectory III. The time interval during which this process lasts is called the time boundary layer. Systems of this kind are called systems with boundary layer.

An estimate of a boundary-layer size t_1 can be obtained from (10.11). It is supposed that an imaging point moving along the trajectory II reaches the trajectory III if an error of coincidence is of the same order μ as the dropped summands in the equations (10.10) and (10.14). Let $v_0+x_0 = O(1)$ and $v(t_1) + x_0 = O(\mu)$. Then from (10.11) it follows that

$$t_1 \sim -\ln \mu. \tag{10.15}$$

It is interesting to evaluate a size of boundary layer t_2 in slow dimensionless time $t = T/T_2$, which appears if $T_* = T_2$. Since $t_2 = t_1/\mu$, then from (10.15) it follows that

$$t_2 \sim -\mu \ln \mu. \tag{10.16}$$

From (10.16) it is obvious that as $\mu \to 0$ the side of boundary layer t_2 tends to zero.

CHAPTER IV

Decomposition of motion in systems with boundary layer

11 Tikhonov theorem

11.1 Introductory considerations

While analyzing the oscillator with high friction in Sec. 10.2 we met a new situation. Two kinds of dimensionless time were spoken about for the identical problem: slow time of slow quasistationary motion for x, and fast time for v inside the boundary layer.

To formalize this circumstance, slow dimensionless time and fast dimensionless time are designated by $t = T/T_2$ and $\tau = T/T_1$, respectively. They are connected with each other by the relation

$$\tau = \frac{t}{\mu}, \qquad \mu = \frac{T_1}{T_2} \ll 1. \tag{11.1}$$

Here a small parameter is denoted by μ, which is traditional for systems with a boundary layer.

An arbitrary dynamic system with fast and slow variables is considered. Its equations, written for fast dimensionless time, have the form (10.1). In (10.1) time is denoted by τ and small parameter by μ:

$$\frac{dy}{d\tau} = \mu Y, \quad \frac{dz}{d\tau} = Z; \quad y(0) = y_0, \quad z(0) = z_0. \tag{11.2}$$

Rewriting (11.2) also for slow time t, which is a traditional consideration for systems with a boundary layer, and using (11.1) leads to:

$$\frac{dy}{dt} = Y, \quad \mu\frac{dz}{dt} = Z; \quad y(0) = y_0, \quad z(0) = z_0. \tag{11.3}$$

At first glance, Eqs. (11.2) and (11.3) belong to different types of systems, that is, regularly and singularly perturbed systems. However, both systems are written for the same singularly perturbed problem. In correspondence with definitions from Sec. 2.3, system (11.3) is singularly perturbed, because it contains small parameters under derivatives and is considered at a finite interval of dimensionless slow time $0 \leq t \leq t' < \infty$. Here $t' = T'/T_2$, and $T' \sim T_2$. System (11.2) is also singularly perturbed, because the time $T' \sim T_2$ corresponds to an asymptotically large interval of fast dimensionless time of this system $0 \leq \tau \leq T'/T_1 = t'/\mu$.

So it can be said that small-parameter singularity for the system (11.3) is equivalent to asymptotically large-interval singularity for (11.2).

But important differences take place for systems with fast phase, usually described by Eqs. (11.2), and systems with boundary layer with their traditional Eqs. (11.3).

For systems with fast phase, intensity of fast motions is kept at the whole interval of motion. Its integral action on slow motion is taken into account by an averaging operation.

For systems with a boundary layer, fast motions are damped sharply inside the boundary layer, and both components of the motion can be investigated separately. The methods of approximate analysis of systems with a boundary layer [7, 25 – 30] are based on these reasons. Such systems form a more restricted class of dynamic systems as compared with systems with fast phase, so the method of their approximate analysis is simpler than the averaging method, that is, requires less tedious calculations.

It should be noted that the idea of analyzing of a problem with two arguments t and τ outside and inside the boundary layer is developed consequently in Lomov [29], where several types of independent arguments are introduced at the overall time interval of the problem.

11.2 Tikhonov theorem

Let us consider a singularly perturbed system of general form

$$\begin{aligned} \frac{dy}{dt} &= Y(y,z,t,\mu); \quad y(0) = y_0, \\ \mu\frac{dz}{dt} &= Z(y,z,t,\mu); \quad z(0) = z_0, \qquad \mu \ll 1, \end{aligned} \tag{11.4}$$

where y and z are n- and m-dimensional vectors, correspondingly.

A degenerate (according to A. N. Tikhonov [25]) system of equations is obtained from (11.4) by setting $\mu = 0$ and excluding z_0:

$$\begin{aligned} \frac{d\overline{y}}{dt} &= Y(\overline{y}, \overline{z}, t, 0); \qquad \overline{y}(0) = y_0, \\ 0 &= Z(\overline{y}, \overline{z}, t, 0). \end{aligned} \tag{11.5}$$

The question of proximity of solutions for the systems (11.4) and (11.5) is posed.

In the following several definitions are formulated.

Definition 1. The root $z = \varphi(y,t)$ of a system of equations

$$Z(y,z,t,0) = 0 \tag{11.6}$$

is called isolated at some restricted range of variables y and t, if the other roots of the system (11.6) do not exist for each fixed value y and t in any small proximity of the root.

Let us introduce a system of equations of zeroth approximation for fast motions. This system is called "adjoined," or the boundary layer system [25]. To get it, transition to the fast time $\tau = t/\mu$ in (11.4) and the condition $\mu = 0$ should be realized.

$$\frac{dz}{d\tau} = Z(y,z,t,0). \tag{11.7}$$

Here slow y and t are considered as parameters. Transition to τ at the right-hand side is not implemented because for any finite t when $\mu \to 0$ it follows that $\tau \to \infty$, and $\mu\tau = t$. Comparison of (11.7) with (10.3) shows that the adjoined system (according to Tikhonov) coincides with the generating system of the Poincaré and the averaging methods.

It could be supposed that semantic nuances of the terms "generating" and "adjoined" are explained by the fact that when new terms became necessary, A. Poincaré worked with regular perturbed systems, whence A. N. Tikhonov analyzed singularly perturbed ones.

According to Sec. 2.2.2 the solution of the system (11.7) exists in fast dimensionless time $\tau = T/T_1$. The interval of this fast motion is called the boundary layer.

Obviously the root $z = \varphi(y, t)$ of Eq. (11.6) defines the stationary point, that is, the equilibrium position of the system (11.7).

Definition 2. The domain of influence of the stationary point $z = \varphi(y, t)$ is called the set of the points z^+ such that solutions of the adjoined system, with initial points z^+, tend to $\varphi(y, t)$ as $t \to \infty$.

The Tikhonov theorem:

Consider:

1. *Functions $Y(y, z, t, \mu)$ and $Z(y, z, t, \mu)$ are analytical with respect to y, z, t, and μ in some domain of the variables' space.*

2. *Equation (11.6) has a root $z = \varphi(y, t)$ in some limited domain D of variables y and t, and this root is isolated.*

3. *Function $Y(y, \varphi(y, t), t, 0)$ is analytical with respect to y and t.*

4. *Initial conditions z_0 are in the domain of influence of the root $z = \varphi(y, t)$ of the system (11.7).*

5. *Stationary points $z = \varphi(y, t)$ of the system (11.7) are asymptotically stable according to Lyapunov for all y and t, for which the root of Eq. (11.6) is defined.*

Then there exist such $\mu_0 > 0$ that the unique solution of the system (11.4) exists for $0 \le \mu \le \mu_0$, and it satisfies the limit equalities:

$$\begin{aligned} \lim_{\mu \to 0} y(t, \mu) &= \overline{y}(t) \text{ for } 0 \le t \le t', \\ \lim_{\mu \to 0} z(t, \mu) &= \overline{z}(t) \text{ for } 0 < t \le t'. \end{aligned} \tag{11.8}$$

Here $0 \le t \le t'$ is a finite time interval that lies inside time interval D, where the unique solution of the system (11.4) exists. The null point does not belong to the domain $0 < t \le t'$ for z from (11.8). It excludes a boundary layer of infinitesimal magnitude where the difference $||z - \overline{z}||$ is large as in (10.16).

Let us consider the sense of Conditions 1 – 5 of the Tikhonov theorem. Condition 1 guarantees existence and uniqueness of the solution of the initial system (11.4) for any finite μ. (For $\mu \to 0$ the problem is not clear because of the singularity $1/\mu$ arising in the equation $dz/dt = Z/\mu$. The theorem states that solution (11.4) exists also for $\mu \to 0$.)

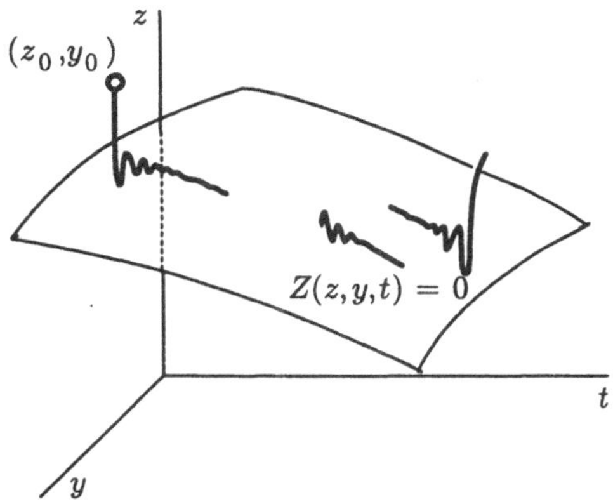

Fig. 12

Condition 2 guarantees the possibility of excluding the variable z in the degenerate system (11.5). Therefore, it is possible to write (11.5) only in slow variables $\overline{y}$, and formulation of Cauchy problem is possible for the degenerate system.

$$\frac{d\overline{y}}{dt} = Y(\overline{y}, \varphi(\overline{y}, t), t, 0), \qquad \overline{y}(0) = y_0. \tag{11.9}$$

Condition 3 guarantees existence and uniqueness of the solution of the degenerate system (11.9). At the space of initial variables y and z according to (11.5), it is placed on the surface $Z(y, z, t, 0) = 0$.

If Condition 4 is satisfied, the trajectory of the initial equation (11.4) at the space (z, t) gets onto the surface $Z(y, z, t, 0) = 0$ during the boundary layer time.

If Condition 5 is satisfied, the trajectory of the initial system is attracted to the surface $Z(y, z, t, 0) = 0$, whereas the motion of the degenerate system evolves on it. It is obvious that Condition 5 guarantees the existence of at least a small domain of influence according to Condition 4. If Condition 5 is not satisfied, the trajectory of the initial system leaves the surface $Z(y, z, t, 0) = 0$, and one cannot speak about the proximity of the solutions of the initial and the degenerate systems. Some situations arising in the process of realization and violation of Conditions 4 and 5 are shown in Fig. 12.

Thus, the most essential condition of the Tikhonov theorem is the requirement of "attraction" of the trajectories of the initial system by the surface $Z(y, z, t, 0) = 0$. For its verification it is natural to apply the theorem about stability of first approximation for an adjoined system (11.7). After substitution $z = \varphi(y, t) + \widetilde{z}$ in (11.7), where $\widetilde{z}$ is a small deviation from the stationary point, the linear equation with respect to $\widetilde{z}$ is

$$\frac{d\widetilde{z}}{d\tau} = \left[\frac{\partial Z_i}{\partial z_j}\right] \widetilde{z}. \tag{11.10}$$

Here $\widetilde{z}$ is an m-dimensional column, and $[\partial Z_i/\partial z_j]$ is a square $m \times m$-dimensional matrix. Elements of the ith row of the matrix are formed by partial derivatives of the ith component of the vector Z for all variables z_j, $j = 1, \ldots, m$. The values of partial derivatives are calculated at the point $z = \varphi(y, t)$; that is, components of the matrix depend on y and t, which are assumed to be constant parameters. So (11.10) is a linear system with constant coefficients.

Remark 1. The Tikhonov theorem states asymptotic proximity (11.8) of solutions of the initial and degenerate systems—without estimation of the order of this proximity. The proximity of the first order of smallness ($\|z - \overline{z}\| = O(\mu), \|y - \overline{y}\| = O(\mu)$) is provided [26] if Condition 5 of the theorem is satisfied due to the first approximation equations (11.10).

Remark 2. Let us consider a dynamic system with a hierarchy of numeric values of the time constants. Let the equation system (2.6) have the form

$$\frac{T_1}{T_*}\frac{dx_1}{dt} = f_1, \ldots, \frac{T_n}{T_*}\frac{dx_n}{dt} = f_n, \tag{11.11}$$

where $T_1 \ll T_2 \ll \ldots \ll T_n$.

For time normalization let us choose $T_* = T_n$. Then (11.11) is transformed into

$$\mu_1 \frac{dx_1}{dt} = f_1, \quad \ldots, \quad \mu_{n-1}\frac{dx_{n-1}}{dt} = f_{n-1}, \quad \frac{dx_n}{dt} = f_n, \tag{11.12}$$

where $\mu_1 \ll \mu_2 \ll \ldots \ll \mu_{n-1} \ll 1$.

If in (11.12) it is set that $\mu_1 = \ldots = \mu_{n-1} = 0$, then the obtained singular system describes the slowest components of the motion for the time scale T_n. It is shown in Tikhonov [25] that to prove the admissibility of the transition to this $(n-1)$–times degenerate system, realization of the conditions of the Tikhonov theorem for sequential degenerations (first for μ_1, then for μ_2, etc.) should be checked.

Indexing in the chain of the hierarchically decreasing small parameters of the system (11.12) is similar to Tikhonov [25]. This order of enumeration is inverse to traditionally accepted in the oscillation theory [24], where characteristic frequencies are enumerated in the order of increasing values, that corresponds to decreasing the related time constants. Sometimes, for example, in case of extraction of the slowest low frequency motions from some spectrum of motions, the latter method of indexing is preferable.

Remark 3. The Tikhonov theorem is formulated for equations such as (11.4) with right-hand sides that depend explicitly on slow time t. In this case, the argument $t = T/T_2$ is changed by the value of a unity order, when initial dimensional time T is changed by the value of the order of the greater time constant T_2. Let the right-hand sides of Eqs. (11.4) explicitly depend on fast time. Then the argument T/T_1 is changed by the value of unity order when T is changed by the value of small time constant of order T_1. Then, for example, the second equation of (11.4), represented in time T, has the form

$$T_1 \frac{dz}{dT} = Z(y, t, \frac{T}{T_1}, \mu),$$

whereas transition to $t = T/T_2$ yields

$$\frac{T_1}{T_2} \frac{dz}{dt} = Z(y, t, \frac{T_2}{T_1} t, \mu),$$

or

$$\mu \frac{dz}{dt} = Z(y, t, \frac{t}{\mu}, \mu).$$

In the last equation the function $Z(y, t, t/\mu, \mu)$ does not satisfy the requirement of the Tikhonov theorem concerned with analyticity with respect to μ.

Combining this remark with Conditions 4 and 5 of the theorem it is seen that the Tikhonov theorem can be applied for investigation of dynamic systems which satisfy the following conditions: fast components of free oscillations of the system have decreasing behavior, and the system is subjected to actions that depend explicitly on time and that are slow.

11.3 Decomposition of motion on an infinite time interval

The conditions of the Tikhonov theorem provide proximity of solutions of the initial and singular systems only on the finite time interval $0 < t \leq t'$. In Kus'mina [30] the proximity is estimated on an asymptotically large time interval as $\mu \to 0$. While investigating the concrete dynamic system for the finite μ, it is natural that fear arises connected with the fact that the time of the system motion is found to be more than these time intervals. In this situation it is interesting to discuss a possibility of application of degenerate equations "with reserve," on the whole infinite time interval.

The behavior of solutions of differential equations with small parameter multiplied by derivatives is studied on the infinite interval in Gradstein [27] and Krasovsky and Klimushev[28]. According to these papers, uniform asymptotic stability of an arbitrary particular solution of a degenerate system is required for the degeneracy on the infinite time interval. This conclusion is the extension of the theorem about stability under permanently acting regular perturbations for the case of singular perturbations. Corresponding theorems in general form are presented by Krasovsky and Klimushev [28].

An analogous extension of the time interval for averaging methods is implemented by Banfi's theorem; the reference to it is presented in Sec. 6.5.

In Gradstein [27] the problem is stated for the particular case. The investigation is reduced to linear equations with a constant coefficient; therefore, it can be completed easily.

The conclusions of the aforementioned I. S. Gradstein paper are presented in the following using the terminology and notations previously introduced.

The system of nonlinear differential equations with small parameter μ multiplied by some derivatives is considered. It is supposed that the right-hand sides of the equations satisfy the requirements mentioned in Tikhonov [25] and Vasil'eva and Butusov [26].

In Gradstein [27] a linearization of this initial system in the proximity of some particular solution of the singular system is performed. It is supposed that coefficients at the linear part of the expansion do not depend on time, but only on μ.

Let us consider two systems of linear differential equations with constant coefficients. One is obtained if linearization of the initial system is fulfilled only with respect to fast variables. This system coincides with the equations obtained by linearization of the adjoined system in the proximity of its stationary points.

The other system is obtained by linearization of the degenerate system of equations.

It is supposed that the real parts of roots of the characteristic equations for both systems are less than -3λ, where λ is an arbitrary positive number. Then for any $\varepsilon > 0$ such $\delta(\varepsilon)$ and $\overline{\mu}$ can be found that the absolute value of the deviation of every variable of the initial system relative to the particular solution of the degenerate system will be less than $\varepsilon \exp(-\lambda t)$ for any t, if initial absolute values of such deviations are less than $\delta(\varepsilon)$, and parameter μ is chosen inside the interval $0 \leq \mu \leq \overline{\mu}$.

The wide class of the systems which satisfy the I. S. Gradstein conditions can be easily indicated.

The system, degenerate equations of which have particular solutions similar to stationary point solutions, is considered. Let this point be asymptotically stable with respect to the first approximation of degenerate equations. Let also the domain of influence of the stationary point coincide with the domain D, where the solution of the degenerate system is determined.

It is assumed that initial equations of the system satisfy the admissibility conditions of the transition to the degenerate equation, specified in Tikhonov [25] and Vasil'eva and Butusov [26]. Then we can assign a value of μ such that it provides the entrance of the solution of the initial system during the boundary layer time to an arbitrary small vicinity of the solution of the degenerate system, and also "together with this solution" during the finite time t' the solution enters into a vicinity of the stationary point of the degenerate system. It can be easily checked that when $t > t'$ the system satisfies all requirements of I. S. Gradstein.

During each stage of the investigation (the stage of entrance of the solution into the δ– vicinity of the equilibrium position and the stage of the motion inside this vicinity during infinite time) the existence of such value of the small parameter $\overline{\mu}$ is proved, which solves the problem at its stage. As the value of the small parameter, which solves the problem as a whole, it can be considered the smallest of these two values.

12 Application of the Tikhonov theorem

In this section it is shown on simple examples that a number of approximate methods, used widely in practice, can be justified by means of the Tikhonov theorem.

12.1 Quasistatic motions of mechanical systems

12.1.1. Let us consider a weakly damped mechanical oscillation system with one degree of freedom, perturbed by the force, slowly changing in time. While describing this class of motion in Sec. 2.1.2(c), the following system of equations was obtained

$$\mu\frac{dv}{dt} = -2\zeta v - x + f(t),$$
$$\mu\frac{dx}{dt} = v, \quad \mu = \frac{T_0}{T_\text{p}} \ll 1. \tag{12.1}$$

The system (12.1) is singularly perturbed. The degenerate system for it is obtained by setting $\mu = 0$:

$$0 = -2\zeta\,\overline{v} - \overline{x} + f(t),$$
$$0 = \overline{v}\,. \tag{12.2}$$

The first equation in (12.2) is the quasistatic equation of the force balance, dimensional analogues of which form the right-hand side of Eq. (12.2). From (12.2) it follows that $\overline{x} = f(t)$, which is the traditional "quasistatic" solution of the system (12.1).

Let us check whether the conditions of the Tikhonov theorem from Sec. 11.2 are realized for (12.1). Conditions 1, 2, and 3 are realized obviously. The adjoined system for (12.1) is constructed by a change $t = \mu\tau$:

$$\frac{dv}{d\tau} = -2\zeta v - x + f(t), \quad \frac{dx}{d\tau} = v. \tag{12.3}$$

Here slow time is considered as parameter, $t = const$. The system (12.3) is linear, so Conditions 4 and 5 of the theorem coincide: the domain of influence of the stationary point for (12.3) is unlimited. Obviously, the stationary point for (12.3) is asymptotically stable, so Conditions 4 and 5 are also satisfied.

12.1.2. Let the friction be "large," as in Sec. 2.1.2 (d). While describing this class of motion, in Sec. 2.1.2 (d) the following system of equations was obtained.

$$\begin{aligned} \mu\frac{dv}{dt} &= -v - x + f(\nu t); \quad v(0) = v_0 = \frac{\dot{x}_0 T}{x_0}, \quad \mu = \frac{T_1}{T_2} \ll 1, \\ \frac{dx}{dt} &= v, \quad x(0) = x_0 \sim 1, \quad \nu = \frac{T_2}{T_{\mathrm{p}}} \sim 1. \end{aligned} \tag{12.4}$$

The singular system for (12.4) has the form

$$0 = -\overline{v} - \overline{x} + f(\nu t), \qquad \frac{d\overline{x}}{dt} = \overline{v}. \tag{12.5}$$

The first equation in (12.5) is the quasistatic equation of the force balance. Excluding the variable v in (12.5) leads to

$$\frac{d\overline{x}}{dt} + \overline{x} = f(\nu t). \tag{12.6}$$

Equation (12.6) describes filtration of the perturbation by an aperiodic element, which has time constant T_2 of the same order, as characterizing time T_{p} of the perturbation.

Here the conditions of the Tikhonov theorem are satisfied obviously, as in Sec. 12.1.1. The only difference is that instead of an equation of the second order (12.3), the adjoined equation is:

$$\frac{dv}{d\tau} + v = -x + f(\nu t) = \text{const}.$$

By analogy with Sec. 10.2, this equation describes fast motions with quasistatic character of velocity.

12.1.3. Let us consider a more difficult "quasistatic" problem connected with the pendulum motion in a strongly viscous medium. The equations of the motion are written in the form:

$$I\frac{d^2\Phi}{dT^2} + R\frac{d\Phi}{dT} + mg\sin\Phi = 0; \quad \Phi(0) = \Phi_0, \quad \frac{d\Phi}{dT}(0) = \Omega_0. \tag{12.7}$$

Here, in addition to the notations from (4.13), the coefficient of the viscous friction is denoted by R. Dividing Eq. (12.7) by R and transforming it to Cauchy form leads to:

$$\begin{aligned} T_1\frac{d\Omega}{dT} &= -\frac{1}{T_2}\sin\Phi - \Omega; \quad \Omega(0) = \Omega_0, \\ \frac{d\Phi}{dT} &= \Omega; \qquad \Phi(0) = \Phi_0. \end{aligned} \tag{12.8}$$

Here $T_2 = R/mgl$, $T_1 = I/R$ are the multipliers with the dimension of time. The normalization of the equations is performed according to the following relations:

$$t = \frac{T}{T_*}, \qquad \varphi = \frac{\Phi}{\Phi_*}, \qquad \omega = \frac{\Omega}{\Omega_*}.$$

The class of the motion with significant angles ($\Phi_* = 1$) in a strongly viscous medium ($T_2 \gg T_1$) is considered. By analogy with Sec. 2.1.2(d) it is assumed that $\Omega_* = \Phi_*/T_2$. For extracting slow components of the motion, it is accepted that $T_* = T_2$. Then (12.8) takes the form:

$$\begin{aligned} \mu\frac{d\omega}{dt} &= -\sin\varphi - \omega; \quad \omega(0) = T_2\Omega_0, \quad \mu = \frac{T_1}{T_2} \ll 1, \\ \frac{d\varphi}{dt} &= \omega; \quad \varphi(0) = \varphi_0 = \Phi_0. \end{aligned} \tag{12.9}$$

The singular system for (12.9) is:

$$0 = -\sin\overline{\varphi} - \overline{\omega}, \quad \frac{d\overline{\varphi}}{dt} = \overline{\omega}; \quad \overline{\varphi}(0) = \varphi_0. \tag{12.10}$$

The first equation in (12.10) is the equation of quasistatic momentum balance.

The system (12.10) is transformed to equation

$$\frac{d\overline{\varphi}}{dt} = -\sin\overline{\varphi}, \quad \overline{\varphi}(0) = \varphi_0, \tag{12.11}$$

which can be simply integrated:

$$\overline{\varphi} = 2\arctan(e^{-t}\tan\frac{\varphi_0}{2}). \tag{12.12}$$

It should be noted that the initial system (12.9) is not integrated in elementary functions.

The requirements of the Tikhonov theorem should be checked. The right-hand sides of Eqs. (12.9) and (12.11) are analytical with respect to φ, ω, which provides satisfaction of Conditions 1 and 3. Condition 2 is satisfied because the equation $-\sin\varphi - \omega = 0$ has an isolated root with respect to ω.

The adjoined system for (12.9) has the form:

$$\frac{d\omega}{d\tau} = -\omega - \sin\varphi, \quad \varphi = \text{const}. \tag{12.13}$$

The equilibrium position of Eq. (12.13), which is linear with respect to ω, is asymptotically stable; therefore, Conditions 4 and 5 of the theorem are also satisfied.

Remark. To understand the obtained result, the general question should be formulated: whether the approximate analysis of the initial system (12.9) is necessary, and how useful is the application of the Tikhonov theorem? The point is that analyzing the finite formula (12.12) (even to obtain diagrams) is too tedious a task to fulfill without a computer. It May be more suitable to obtain graphic relations by means of numerical integration of the initial (for (12.12)) degenerate equations (12.11). And if so, is it perhaps better to integrate Eqs. (12.9) on the computer without additional difficulties?

The advantages of the transition from (12.9) to (12.11) are not confined to reducing the number of parameters and decreasing the system order. The main simplification of the problem is in the fact that numerical integration of the initial system (12.9) is harder than the degenerate system (12.11). The integration step of (12.9) is defined by the scale T_1 of fast motions, and the system is integrated over a large time interval with the time scale T_2. In the theory of numerical methods such systems are called "stiff"; their numerical integration causes significant difficulties. The degenerate system is written for the one time scale, and its numerical analysis can be performed by standard methods.

12.2   The method of "frozen coefficients"

The possibility of method verification is illustrated in the example of the one-dimensional problem connected with the towing of a solid in a viscous medium on a cable of variable length.

The attaching point of the cable with the tow is denoted by O; the attaching point of the cable with the solid is denoted by A. The length of the cable is $OA = L + X$, where L is the nominal length of the undistorted cable, and X is the deformation. It is assumed that the nominal length of the cable is changed during towing: $L = L(T/T_{\mathrm{p}})$, where T_{p} characterizes the time of the length change. The motion of the point O with constant velocity is given as $V_O = $ const. The absolute velocity of the point A is:

$$V_A = V_O + \frac{d}{dt}OA = V_O + \frac{dL}{dT} + \frac{dX}{dT}. \qquad (12.14)$$

The equation of the solid motion can be written as:

$$M\frac{dV_A}{dT} = F_1 + F_2. \qquad (12.15)$$

Here F_1 is the force of elasticity proportional to the relative deformation X/L of the cable; F_2 is the force of friction proportional to the absolute velocity of the solid:

$$F_1 = -C(X/L), \qquad F_2 = -RV_A; \qquad (12.16)$$

and C, R are the coefficients of the proportion. Substituting (12,14) and (12.16) into (12.15) yields:

$$M\frac{d^2X}{dT^2} = -R\frac{dX}{dT} - K\left(\frac{T}{T_{\mathrm{p}}}\right)X + F\left(\frac{T}{T_{\mathrm{p}}}\right), \qquad (12.17)$$

where

$$K\left(\frac{T}{T_{\mathrm{p}}}\right) = C/L\left(\frac{T}{T_{\mathrm{p}}}\right), \qquad F\left(\frac{T}{T_{\mathrm{p}}}\right) = -R\left(V_O + \frac{dL}{dT}\right) - M\frac{d^2L}{dT^2}.$$

By analogy with Sec. 12.1.1, it is assumed that the friction does not change the oscillation behavior of the system (12.17). A variable $V = dX/dT$ is introduced, and (12.17) is normalized according to:

$$t = \frac{T}{T_*}, \quad x = \frac{X}{X_*}, \quad v = \frac{V}{V_*}, \quad k = \frac{K}{K_*}, \quad f = \frac{F}{F_*}.$$

K_* and F_* are chosen to be equal to maximum values of the parameters K and F on the considered time interval of the motion. The following notations are introduced: $T_0^2 = M/K_*$, $2\zeta T_0 = R/K_*$. The class of motion is considered which satisfies the condition $T_0 \ll T_{\mathrm{p}}$. As in Sec. 12.1.1, it is chosen that:

$$X_* = F_*/K_*, \quad V_* = X_*/T_0, \quad T_* = T_{\mathrm{p}}.$$

Thus (12.17) is transformed into:

$$\mu\frac{dv}{dt} = -2\zeta v - k(t)x + f(t),$$
$$\mu\frac{dx}{dt} = v. \qquad (12.18)$$

A degenerate system for (12.18) is:

$$0 = -2\zeta\bar{v} - k(t)\bar{x} + f(t),$$
$$0 = \bar{v}. \qquad (12.19)$$

By its form the system (12.19) coincides with (12.2), but in (12.19) the coefficient $k(t)$ depends on time. From (12.19) the quasistatic solution is obtained, which determines the slow component of the deformation subject to slow changing of the nominal cable length.

The adjoined system for (12.18) has the form:

$$\frac{dv}{d\tau} = -2\zeta v - k(t)x + f(t), \qquad \frac{dx}{d\tau} = v. \qquad (12.20)$$

Here the slow time is considered as a parameter: $t = \mathrm{const}$; hence, the results of the analysis of Eqs. (12.20) are the same as in Sec. 12.1.1.

Thus, while studying fast high frequency components of motion described by the adjoined system (12.20), the slow changing in time coefficients of the system are assumed to be constant.

This method is usually called the method of "frozen" coefficients.

12.3 The limit model for a double pendulum of high stiffness

A plane double pendulum consisting of two solids is considered (Fig. 13).

The solid 2, turning about the joint O_2, contains a circular cavity. Circular solid 1, joined with the solid 2 at the point O_1, is placed into the cavity. For simplicity it is assumed that the center of the mass of the solid 1 coincides with the point O_1. The solids of the system are influenced by the weight force and the moments of the elasticity and friction forces with large coefficients of proportion with respect to relative rotation angles and relative angular velocities of the solids. The solid 2 moves in the medium with viscous friction.

The equations of the system motion are written in the form of equations of angular momentum for the system as a whole relative to the point O_2 and for the solid 1 relative to the point O_1:

$$\begin{gathered} \frac{d}{dT}(I_1\Omega_1 + I_2\Omega_2) = -Pl\sin\Phi_2 - R_2\Omega_2, \\ \frac{d}{dT}I_1\Omega_1 = -K_1(\Phi_1 - \Phi_2) - R_1(\Omega_1 - \Omega_2), \\ \frac{d\Phi_1}{dT} = \Omega_1, \qquad \frac{d\Phi_2}{dT} = \Omega_2. \end{gathered} \tag{12.21}$$

Here, in addition to notations of Eqs. (4.13), P denotes the total weight of the system; l is the distance between the point O_2 and the center of the mass of the system; R_2 is a coefficient of the external friction moment for the system; and K_1, R_1 are coefficients of stiffness and friction of the solids'

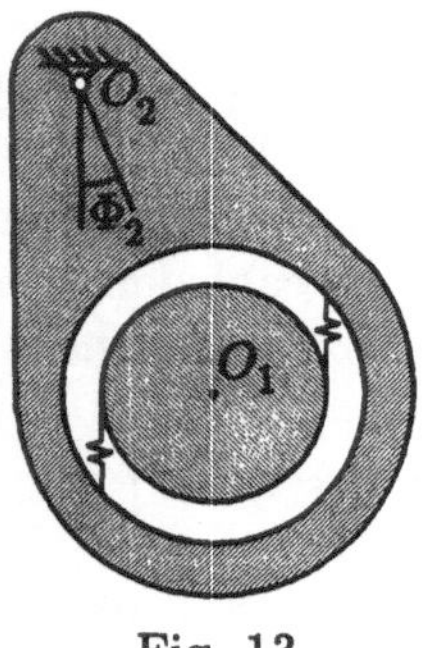

Fig. 13

interaction moments. The indices of notations correspond to the numbers of the solids.

Now in (12.21) the set of phase variables Φ_1, Φ_2, Ω_1, Ω_2 is changed by the set Φ_2, Ω_2, Δ, U, including the variables $\Delta = \Phi_1 - \Phi_2$, $U = \Omega_1 - \Omega_2$, which are responsible for stiff interactions. Therefore,

$$\begin{gathered} I_2 \frac{d\Omega_2}{dT} = -Pl \sin \Phi_2 - R_2 \Omega_2 + K_1 \Delta + R_1 U, \\ I_1 \frac{dU}{dT} = \frac{I_1}{I_2}(Pl \sin \Phi_2 + R_2 \Omega_2) - (1 - \frac{I_1}{I_2})(K_1 \Delta + R_1 U), \\ \frac{d\Phi_2}{dT} = \Omega_2, \qquad \frac{d\Delta}{dT} = U. \end{gathered} \tag{12.22}$$

The change to dimensionless normalized values in (12.22) is carried out:

$$t = \frac{T}{T_*},\ i_1 = \frac{I_1}{I_*},\ i_2 = \frac{I_2}{I_*},\ \varphi_2 = \frac{\Phi_2}{\Phi_*},\ \delta = \frac{\Delta}{\Delta_*},\ \omega_2 = \frac{\Omega_2}{\Omega_*},\ u = \frac{U}{U_*}.$$

The class of motion with significant deviations of the solid 2: $\Phi_* = 1$ is considered. The moments of inertia of the solids have the same order: $I_* = I_2$, and stiffness of the elasticity forces significantly exceeds relaxation coefficient $K_2 = Pl$ due to the pendulum: $K_2 \ll K_1$. Partial time constants of the system should be estimated. The time constant t of the slow oscillation of the system due to pendulum properties is estimated by the relation $T_2^2 = I_2/K_2$. Time constant T_1 of the fast oscillations of the solid 1 due to elasticity can be estimated in the following way: $T_1^2 = I_1/K_1$. If $K_2 \ll K_1$, then $\mu = T_1/T_2 \ll 1$. By analogy with Sec. 4.2.2, characterizing partial angular velocities of the system with respect to the variables Ω_2, U are estimated by the relations $\Omega_* = \Phi_*/T_2$, $U_* = \Delta_*/T_1$. The pendulum moments and the moments of the elastic interaction forces are assumed to be quantities of the same order: $K_1 \Delta_* = K_2$. The quantity of the order of time constant of slow partial oscillations $T_* = T_2$ is chosen to be the characterizing time. After this normalization, (12.22) takes the form:

$$\begin{gathered} \frac{d\omega_2}{dt} = -\sin \varphi_2 - 2\zeta_2 \omega_2 + \delta + 2\zeta_1 u, \quad \frac{d\varphi_2}{dt} = \omega_2, \\ \mu \frac{du}{dt} = i_1(\sin \varphi_2 + 2\zeta_2 \omega_2) - (1 + i_1)(\delta + 2\zeta_1 u), \quad \mu \frac{d\delta}{dt} = u. \end{gathered} \tag{12.23}$$

Here all variables φ_2, ω_2, δ, u have the values of unity order, $i_1 = I_1/I_2$; and ζ_1, ζ_2 are dimensionless damping coefficients of the first and second partial oscillation elements.

Degenerate equations are obtained from (12.23) by setting $\mu = 0$:

$$\begin{gathered} \frac{d\overline{\omega}_2}{dt} = -\sin \overline{\varphi}_2 - 2\zeta_2 \overline{\omega}_2 + \overline{\delta} + 2\zeta_1 \overline{u}, \qquad \frac{d\overline{\varphi}_2}{dt} = \overline{\omega}_2, \\ 0 = i_1(\sin \overline{\varphi}_2 + 2\zeta_2 \overline{\omega}_2) - (1 + i_1)(\overline{\delta} + 2\zeta_1 \overline{u}), \qquad \overline{u} = 0. \end{gathered} \tag{12.24}$$

Excluding $\overline{\delta}$ and $\overline{\omega}_2$from (12.24) yields

$$(1+i_1)\frac{d^2\overline{\varphi}_2}{dt^2}+2\zeta_2\frac{d\overline{\varphi}_2}{dt}+\sin\overline{\varphi}_2=0. \qquad (12.25)$$

It is obvious that after transferring to initial dimensional designations Eq. (12.25) coincides with the equation of oscillation of the absolutely rigid physical pendulum. The moment of inertia of the pendulum in the limit model (12.25) is equal to the sum of inertia moments of the initial solids.

It is readily seen that the limit transition from (12.23) to (12.25) can be treated as an imposition of the holonomic constraint $\Phi_1=\Phi_2$. Now the correctness of this transition is considered according to Sec. 11.2.

Condition 2 of the theorem from Sec. 11.2 is satisfied because the root $u=0$, $\delta=i_1(\sin\varphi_2+2\zeta_2\omega_2)/(1+i_1)$ of the finite equations from (12.24) is unique and isolated. Found value δ is equal to the normalized value of the reaction of the holonomic constraint.

To check Condition 5, adjoined equations for (12.23) should be written. Transition to fast time $\tau=t/\mu$ in (12.23) leads to:

$$\begin{aligned}&\frac{du}{d\tau}+2\zeta_1(1+i_1)u+(1+i_1)\delta+i_1(\sin\varphi_2+2\zeta_2\omega_2)=0,\\&\frac{d\delta}{d\tau}=u;\qquad \omega_2,\varphi_2=\text{const}.\end{aligned} \qquad (12.26)$$

It is obvious that the equilibrium positions of the linear system (12.26) are stable asymptotically.

The additional Gradstein condition from Sec. 11.3 concerned with infinity of the time interval is also satisfied, because the particular solutions of Eqs. (12.24) or (12.25) are, in general, asymptotically stable relative to initial perturbation.

The exceptions are separatrixes, that is, such trajectories of system (12.25) that are terminated at the upper equilibrium position as $t\to\infty$. Obviously, they are not stable with respect to small initial perturbation, and for them the Gradstein condition of admissibility of passage to the limit on the infinite time interval is not satisfied. The mechanical sense of this exception is discussed in the following.

The high frequency transient with respect to the fast variables decreases in the boundary layer. However, at the end of the boundary layer it can lead the initial system (12.23) to different sides of the separatrix, The side depends on the initial conditions for fast variables u, δ, which are not taken into account in the limit model (12.25). These small differences from the separatrix lead to significant deviation from it when $t\to\infty$ (Fig. 14).

Thus, after taking into account the aforementioned exception, the limit transition from (12.23) to (12.24) is correct for an infinite time interval.

Now the case is considered for which limit model (12.25) is conservative ($\zeta_2=0$). The additional Gradstein condition in this case is not satisfied

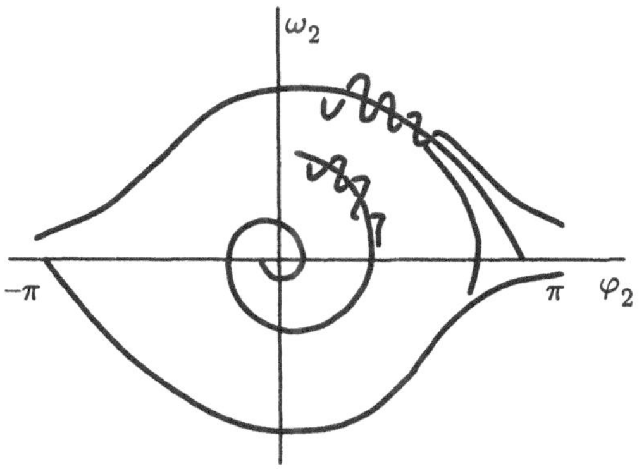

Fig. 14

for all trajectories of this system; therefore, the correctness of the model (12.25) application for the infinite time interval is not provided. Let us consider the mechanical sense of this constraint. The initial conditions of the limit model (12.25) according to Tikhonov [25] and Vasil'eva and Butusov [26] coincide with those for slow variables φ_2, ω_2 of the initial model (12.23). After damping of the transient with respect to fast variables in the boundary layer, the values of the slow variables of the initial system (12.23) will differ from their initial values by the value of order μ [26]. Due to nonisochronism of the system these small differences will lead to finite deviations of the solutions for the initial and limit models when $t \to \infty$.

The differences in the behavior of the initial and degenerate systems are revealed especially easily for small oscillations. The solution of the system (12.25) with $\zeta_2 = 0$ is undamped. The solution of the system (12.23) with $\zeta_2 = 0$ and arbitrarily small $\mu \neq 0$ damps asymptotically to zero. It can be easily proved, for example, with the help of the Gurvitz criterion.

It should be noted that the previously mentioned conditions are sufficient [25–28]. It should be noted also that known varieties of the methods of motion decomposition differ by their formalism and applicability conditions. Refinement and weakening of the requirements, providing validity of the transition under consideration, cannot be excluded.

So the delicate conclusion can be made: violation of the Gradstein conditions in conservative problems on the infinite time interval causes uncertainty, connected with correctness of application of the rigid body models and holonomic constraints. The axiomatic character of such models for similar problems seems to be subject to some doubt.

Finally it should be noted that a considered approach leads to an interesting, carefully worded formal explanation of the noncorrectness of the time inversion in limit models. Macroprocesses, observed in nature, from this point of view are realized only due to asymptotic stability of the high frequency motions on various microlevels. The inverse movement of the time could lead to explosive growth of all fast motions. The events observed in this case will have nothing to do with phenomena, which are described by the limit model for time inversion.

12.4 Relaxation oscillations of the valve generator

For the values of the parameters, defined in Sec. 7.2, harmonic oscillations (7.27) are excited in the valve generator (vacuum tube oscillator). The other range of the numeric values of the parameters, which causes so-called relaxation oscillations with discontinuities with respect to some phase variables, is considered in this section.

From the methodological point of view this problem is interesting because for its solution all conditions of the Tikhonov theorem are essential. For previous mechanical problems these conditions were satisfied almost trivially.

The equations of the generator (7.22) from Sec. 7.2 are considered. The variable $V = dI/dT$ is introduced, and normalization $i = I/I_*$, $v = V/V_*$, $t = T/T_*$ is fulfilled.

Thus Eqs. (12.27) take the form:

$$\begin{aligned} \frac{T_1^2 V_*}{T_*}\frac{dv}{dt} &= T_2 V_* v - A V_*^3 v^3 - I_* i, \\ \frac{I_*}{T_*}\frac{di}{dt} &= V_* v. \end{aligned} \tag{12.27}$$

The characterizing values are connected by the relation: $V_* = I_*/T_*$. Equations (12.27) are transformed into

$$\frac{T_1^2}{T_*^2}\frac{dv}{dt} = \frac{T_2}{T_*}v - A\frac{I_*^2}{T_*^3}v^3 - i, \quad \frac{di}{dt} = v. \tag{12.28}$$

In contrast to Sec. 7.2, the system with parameters satisfies the condition $T_1 \ll T_2$. Two independent dimensional units T_* and I_* are determined by the relations:

$$T_* = T_2, \qquad \frac{AI_*^2}{T_2^3} = 1. \tag{12.29}$$

Therefore, the equations of motion can be written in their final form:

$$\begin{aligned} \mu\frac{dv}{dt} &= v - v^3 - i, \qquad \mu = \frac{T_1^2}{T_2^2} \ll 1, \\ \frac{di}{dt} &= v. \end{aligned} \tag{12.30}$$

A degenerate system for (12.30) is:

$$0 = \bar{v} - \bar{v}^3 - i, \qquad \frac{d\bar{i}}{dt} = \bar{v}. \tag{12.31}$$

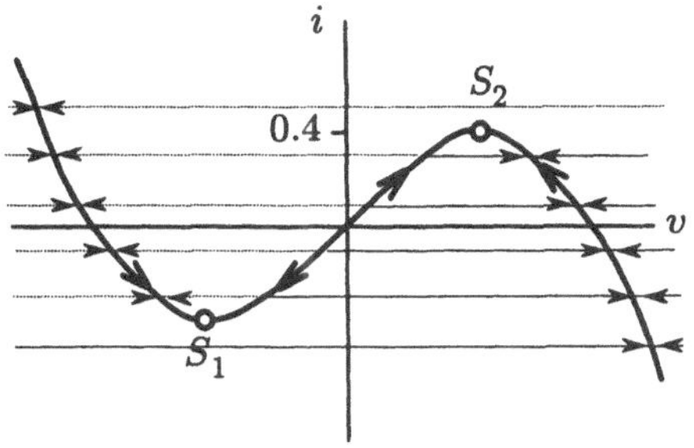

Fig. 15

From (12.31) it is easily seen that the solution of the singular system is developed in the space of the variables i, v along the curve $i = v - v^3$ (analogue of the surface $Z(y, z, t, 0) = 0$ from Sec. 11.2.

From (12.31) it follows that the sign of $d\bar{i}/dt$ coincides with the sign of $\bar{v}$. On Fig. 15 the direction of the motion along the branches of the curve $i(v)$ is designated by arrows. It is readily seen that the system certainly comes into one of the positions corresponding to the minimum S_1 and maximum S_2 of the curve. The degenerate system cannot answer the question of how the process will develop further. The system cannot remain in S_1, S_2, because in these points $di/dt \neq 0$.

The system behavior with respect to fast motion is considered. The adjoined system has the form:

$$\frac{dv}{d\tau} = v - v^3 - i, \qquad i = \text{const}. \tag{12.32}$$

The stationary points of the system (12.32) are determined by the roots of the equation

$$0 = v - v^3 - i. \tag{12.33}$$

Depending on the value of i, Eq. (12.33) has either one or three roots with respect to v. On Fig. 15 they are determined as the intersection of the graph with straight line $i = \text{const}$. All roots are isolated, in addition to the points S_1 and S_2, where two roots coincide.

In the following the stability of the isolated stationary points $v = v^0$ is analyzed. The change $v = v^0 + \Delta v$ in (12.32) yields:

$$\frac{d(v^0 + \Delta v)}{d\tau} = g(v^0 + \Delta v) - i, \qquad \text{where } g(v) = v - v^3.$$

The equations of the first approximation are:

$$\frac{d\Delta v}{d\tau} = g(v^0) + \frac{dg}{dv}(v^0)\Delta v - i.$$

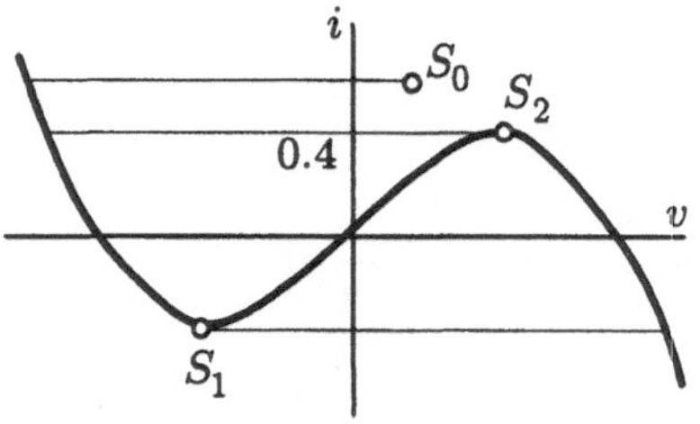

Fig. 16

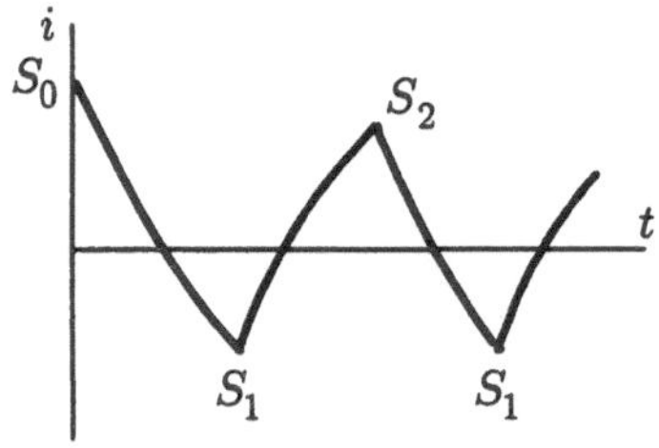

Fig. 17

Since $g(v^0) - i = 0$, then

$$\frac{d\Delta v}{d\tau} = \frac{dg}{dv}(v^0)\Delta v. \tag{12.34}$$

It follows from (12.34) that for $dg/dv > 0$, that is, for the part of the curve on Fig. 15 lying between the points S_1 and S_2, the stationary points are unstable.

Now the domain of influence of the stationary points should be found. For this purpose, the sign of $dv/d\tau$ while the plane of variables i, v intersects the straight line i = const is analyzed. The sign of $dv/d\tau$ is determined from (12.32) by the sign of the expression $v - v^3 - i$. When $v \to +\infty$, we have $v - v^3 - i < 0$; when $v \to -\infty$, we have $v - v^3 - i > 0$; for any intersection by the curve $v - v^3 - i = 0$ the sign of the analyzed expression is changed. Such analysis makes it possible to construct the domain of influence of the curve $v - v^3 - i = 0$, as shown in Fig. 15. The uninterrupted lines are used to signify the domain of influence of the right branch of the curve; the dotted line is used for the left one.

Now the phase trajectory of the system motion is constructed on the plane of variables i, v, with the origin in an arbitrary initial point S_0 (see Fig. 16). The point S_0 is placed in the domain of influence of the left branch of the curve. From S_0 the system "fastly" with the velocity $dv/dt \sim 1/\mu$ arrives at the left branch. Along this branch the system "slowly" goes up to point S_1, and performs the "fast" jump to the right branch. The system moves along it up to S_2, where it falls to the left branch. In the foolowing, the process is performed along the limit cycle. The development of the process with respect to the variable i in time t is shown in Fig. 17. Attention should be paid to the breaks of the graph with jumps of the derivative, corresponding to the points S_1 and S_2.

The amplitude of the oscillations $i_{\max}$ for the variable i, as seen in Fig. 16, equals 0.4. After transition to the initial dimensional variables with the help of (12.29), it can be found that the amplitude of the oscillations $I_{\max}$ with respect to the variable I is equal to

$$I_{\max} = I_* i_{\max} = 0.4\sqrt{T_2^3/A}.$$

13 Asymptotic expansion of solutions for systems with a boundary layer

13.1 Algorithm of expansion

13.1.1 The Tikhonov theorem from Sec. 11.2 extracts only a trivial approximation with respect to μ for slow components of the motion for the system with a boundary layer. Next, following Vasil'eva and Butusov [26], the procedure of construction of expansion is presented, which makes it possible to obtain an approximate solution with arbitrary required asymptotic accuracy. Notations used in Vasil'eva and Butusov [26] are changed partially to provide unique notation throughout this book.

The equations of the system with a boundary layer are considered:

$$\begin{aligned} \frac{dy}{dt} &= Y(y,z,t,\mu), \qquad y(0) = y_0, \\ \mu\frac{dz}{dt} &= Z(y,z,t,\mu), \qquad z(0) = z_0, \quad \mu \ll 1. \end{aligned} \tag{13.1}$$

It is assumed that (13.1) satisfies all requirements of the Tikhonov theorem from Sec. 11.2. Let us reinforce Condition 5 of the asymptotic stability of the stationary points of the adjoined system by the requirements of asymptotic stability for the first approximation.

By analogy with (3.3) and (6.12) for the regularly perturbed systems and the systems with fast phases, the solution of the system (13.1) is searched as the sum of the main slow part $\overline{y}$, $\overline{z}$ and fast additions $\widetilde{y}$, $\widetilde{z}$:

$$\begin{aligned} y &= \overline{y} + \widetilde{y}, \\ z &= \overline{z} + \widetilde{z}. \end{aligned} \tag{13.2}$$

Definition of the redundant substitution (13.2) should be completed.

1. As opposed to the systems with fast phase, here the additions $\widetilde{y}$, $\widetilde{z}$ cannot be considered as small due to "large" rejections in the boundary layer. Therefore, the conditions similar to (6.13) are not imposed on $\widetilde{y}$, $\widetilde{z}$ in (13.2).

2. The processes for $\widetilde{y}$, $\widetilde{z}$ in the systems with a boundary layer occur inside the boundary layer with respect to the fast time τ; the processes for $\overline{y}$, $\overline{z}$ outside the boundary layer take place with respect to slow time t. So, as opposed to (6.14), for (13.2) it is assumed that:

$$\begin{aligned} &\overline{y} = \overline{y}(t,\mu), \quad \overline{z} = \overline{z}(t,\mu); \\ &\widetilde{y} = \widetilde{y}(\tau,\mu), \quad \widetilde{z} = \widetilde{z}(\tau,\mu), \qquad \tau = t/\mu. \end{aligned} \tag{13.3}$$

3. Let us require the motion with respect to $\widetilde{y}$, $\widetilde{z}$ to be finished inside the boundary layer:

$$\|\widetilde{y}(\tau)\|, \quad \|\widetilde{z}(\tau)\| \to 0 \text{ as } \tau \to \infty. \tag{13.4}$$

The solution of the system (13.1) is sought in the form of power expansions with respect to μ separately for the components of the motion (13.3) outside and inside the boundary layer. Therefore, outside the boundary layer

$$\begin{aligned} \overline{y}(t,\mu) &= \overline{y}^{(0)}(t) + \mu\overline{y}^{(1)}(t) + \dots, \\ \overline{z}(t,\mu) &= \overline{z}^{(0)}(t) + \mu\overline{z}^{(1)}(t) + \dots. \end{aligned} \tag{13.5}$$

Inside the boundary layer

$$\begin{aligned} \widetilde{y}(\tau,\mu) &= \widetilde{y}^{(0)}(\tau) + \mu\widetilde{y}^{(1)}(\tau) + \dots, \\ \widetilde{z}(\tau,\mu) &= \widetilde{z}^{(0)}(\tau) + \mu\widetilde{z}^{(1)}(\tau) + \dots. \end{aligned} \tag{13.6}$$

Unknown coefficients $\overline{y}^{(0)}(t)$, $\overline{z}^{(0)}(t), \dots$, $\widetilde{y}^{(0)}(t)$, $\widetilde{z}^{(0)}(t), \dots$, of the expansions (13.5) and (13.6) should be defined with the help of substitution of (13.2), (13.5), and (13.6) into Eqs. (13.1) and setting equal the summands for the same powers of μ. The summands, depending on t and τ, are set equal separately.

The right-hand sides of Eqs. (13.1) should be decomposed beforehand into the summands by the identical transformation. One of the summands must depend only on functions of the argument t, and other of τ:

$$
\begin{aligned}
Y &\equiv \overline{Y} + \widetilde{Y}, \qquad Z \equiv \overline{Z} + \widetilde{Z}, \\
\overline{Y} &\equiv Y(\overline{y}(t,\mu), \overline{z}(t,\mu), t, \mu), \\
\overline{Z} &\equiv Z(\overline{y}(t,\mu), \overline{z}(t,\mu), t, \mu), \\
\widetilde{Y} &\equiv Y(\overline{y}(\mu\tau,\mu) + \widetilde{y}(\tau,\mu), \overline{z}(\mu\tau,\mu) + \widetilde{z}(\tau,\mu), \mu\tau, \mu) \\
&\quad -Y(\overline{y}(\mu\tau,\mu), \overline{z}(\mu\tau,\mu), \mu\tau, \mu), \\
\widetilde{Z} &\equiv Z(\overline{y}(\mu\tau,\mu) + \widetilde{y}(\tau,\mu), \overline{z}(\mu\tau,\mu) + \widetilde{z}(\tau,\mu), \mu\tau, \mu) \\
&\quad -Z(\overline{y}(\mu\tau,\mu), \overline{z}(\mu\tau,\mu), \mu\tau, \mu).
\end{aligned}
\tag{13.7}
$$

It should be noted that after differentiation of the summand $\widetilde{y}(\tau,\mu)$ with respect to t, the multiplier $1/\mu$ appears in the first equation of (13.1). So it is useful to first multiply the equation by μ.

Then

$$
\begin{aligned}
\mu\frac{d\overline{y}}{dt} + \frac{d\widetilde{y}}{d\tau} &= \mu(\overline{Y} + \widetilde{Y}), \\
\mu\frac{d\overline{z}}{dt} + \frac{d\widetilde{z}}{d\tau} &= \overline{Z} + \widetilde{Z}.
\end{aligned}
\tag{13.8}
$$

Differential equations, which will be obtained in the following for the unknown coefficients of the expansions (13.5) and (13.6), will require determination of the initial conditions with respect to these variables. The conditions are determined by substitution of initial conditions y_0, z_0 of the initial problem (13.1) into (13.2), (13.5) and (13.6) for the initial moment of time:

$$
\begin{aligned}
y_0 &= \overline{y}^{(0)}(0) + \mu\overline{y}^{(1)}(0) + \ldots + \widetilde{y}^{(0)}(0) + \mu\widetilde{y}^{(1)}(0) + \ldots, \\
z_0 &= \overline{z}^{(0)}(0) + \mu\overline{z}^{(1)}(0) + \ldots + \widetilde{z}^{(0)}(0) + \mu\widetilde{z}^{(1)}(0) + \ldots.
\end{aligned}
\tag{13.9}
$$

After substitution of (13.2), and (13.5) through (13.7) into (13.8), the equations of zeroth approximation with respect to μ can be written.

Outside the boundary layer the summands, depending on t, give:

$$
\begin{aligned}
\frac{d\overline{y}^{(0)}}{dt} &= Y(\overline{y}^{(0)}, \overline{z}^{(0)}, t, 0), \\
0 &= Z(\overline{y}^{(0)}, \overline{z}^{(0)}, t, 0).
\end{aligned}
\tag{13.10}
$$

Inside the boundary layer

$$\begin{aligned} \frac{d\widetilde{y}^{(0)}}{d\tau} &= 0, \\ \frac{d\widetilde{z}^{(0)}}{d\tau} &= Z(\overline{y}^{(0)}(0) + \widetilde{y}^{(0)}, \overline{z}^{(0)}(0) + \widetilde{z}^{(0)}, 0, 0) \\ &\quad - Z(\overline{y}^{(0)}(0), \overline{z}^{(0)}(0), 0, 0). \end{aligned} \tag{13.11}$$

The initial conditions of systems (13.10) and (13.11) can be obtained from (13.9) by setting equal the components of zeroth order with respect to μ:

$$\begin{aligned} y_0 &= \overline{y}^{(0)}(0) + \widetilde{y}^{(0)}(0), \\ z_0 &= \overline{z}^{(0)}(0) + \widetilde{z}^{(0)}(0). \end{aligned} \tag{13.12}$$

Equations (13.10) and (13.11) form the system of differential equations of the third order. The four initial values $\overline{y}^{(0)}(0)$, $\overline{z}^{(0)}(0)$, $\widetilde{y}^{(0)}(0)$, $\widetilde{z}^{(0)}(0)$ in (13.12) are connected by one finite relation, that is, by the second equation of (13.10). It is impossible to define three remaining independent values from the two equations (13.12). Let us complete the problem (13.4) definition, demanding that outside the boundary layer the solution must be defined only by the summands $\overline{y}$, $\overline{z}$. In conformity to zeroth approximation it means that $\widetilde{y}^{(0)}(\tau)$, $\widetilde{z}^{(0)}(\tau) \to 0$ as $\tau \to \infty$. The condition $\widetilde{z} \to 0$ is satisfied due to the asymptotic stability of the stationary points of the adjoined system. Hence, the condition which completes the definition of the problem is:

$$\widetilde{y}^{(0)}(\tau) \to 0 \quad \text{at} \quad \tau \to \infty. \tag{13.13}$$

Then from the first equation of (13.11) it follows that $\widetilde{y}^{(0)}(\tau) \equiv 0$; that is, $\widetilde{y}^{(0)}(0)=0$. From the first equation of (13.12) it is obtained that $y_0 = \overline{y}^{(0)}(0)$. Then from the second equation of (13.10) we have $\overline{z}^{(0)}(0) = \varphi(y_0, 0, 0)$, where $\varphi(y, 0, 0)$ is the root of the equation $Z(\overline{z}^{(0)}, y_0, 0, 0) = 0$. Finally, from the second equation of (13.12) we obtain $\widetilde{z}^{(0)}(0) = z_0 - \overline{z}^{(0)}(0) = z_0 - \varphi(y_0, 0, 0)$.

Thus, the equations of zeroth approximation are the following.

$$\begin{aligned} \frac{d\overline{y}^{(0)}}{dt} &= Y(\overline{y}^{(0)}, \overline{z}^{(0)}, t, 0), \quad \overline{y}^{(0)}(0) = y_0, \\ 0 &= Z(\overline{y}^{(0)}, \overline{z}^{(0)}, t, 0), \end{aligned} \tag{13.14}$$

$$\begin{aligned} \frac{d\widetilde{z}^{(0)}}{dt} &= Z(y_0, \varphi(y_0) + \widetilde{z}^{(0)}, 0, 0), \\ \widetilde{z}^{(0)}(0) &= z_0 - \varphi(y_0). \end{aligned} \tag{13.15}$$

Equations (13.14) coincide with the equations of the system, which is degenerate according to Tikhonov. Equation (13.15) is obtained from the adjoined system by the substitution $z = \varphi(y_0) + \widetilde{z}^{(0)}$.

By collecting in (13.8) the components with μ, the equations of first approximation are obtained.

Outside the boundary layer

$$\begin{aligned} \frac{d\overline{y}^{(1)}}{dt} &= \left[\frac{\partial Y_i}{\partial y_j}\right] \overline{y}^{(1)} + \left[\frac{\partial Y_i}{\partial z_j}\right] \overline{z}^{(1)} + \left[\frac{\partial Y_i}{\partial \mu}\right], \\ \frac{d\overline{z}^{(0)}}{dt} &= \left[\frac{\partial Z_i}{\partial y_j}\right] \overline{y}^{(1)} + \left[\frac{\partial Z_i}{\partial z_j}\right] \overline{z}^{(1)} + \left[\frac{\partial Z_i}{\partial \mu}\right]. \end{aligned} \tag{13.16}$$

The elements of the matrices $[\partial Y_i/\partial y_j], \ldots$ are calculated here in the point $\overline{y}^{(0)}(t),\ \overline{z}^{(0)}(t)$.

The equations inside the boundary layer are:

$$\begin{aligned} \frac{d\widetilde{y}^{(1)}}{d\tau} &= Y(\overline{y}^{(0)}(0) + \widetilde{y}^{(0)}, \overline{z}^{(0)}(0) + \widetilde{z}^{(0)}, 0, 0) \\ &\quad - Y(\overline{y}^{(0)}(0), \overline{z}^{(0)}(0), 0, 0) \equiv \widetilde{Y}^{(0)}(\tau), \\ \frac{d\widetilde{z}^{(1)}}{dt} &= \left[\frac{\partial Z_i}{\partial y_j}\right] \widetilde{y}^{(1)} + \left[\frac{\partial Z_i}{\partial z_j}\right] \widetilde{z}^{(1)} + G_1(\tau). \end{aligned} \tag{13.17}$$

Here the elements of the matrices $[\partial Y_i/\partial y_j], \ldots$ are calculated in the point $\overline{y}^{(0)}(0) + \widetilde{y}^{(0)}(\tau),\ \overline{z}^{(0)}(0) + \widetilde{z}^{(0)}(\tau)$, and the vector G_1 is calculated according to (13.7) and depends only on τ.

The initial conditions for the systems (13.16) and (13.17) can be obtained by setting equal the summands of the first order with respect to μ in (13.9):

$$\begin{aligned} 0 &= \overline{y}^{(1)}(0) + \widetilde{y}^{(1)}(0), \\ 0 &= \overline{z}^{(1)}(0) + \widetilde{z}^{(1)}(0). \end{aligned} \tag{13.18}$$

As in (13.13), definition of the problem is completed by the condition

$$\widetilde{y}^{(1)}(\tau) \to 0 \text{ as } \tau \to \infty. \tag{13.19}$$

Integration of the first equation of (13.17) leads to:

$$\widetilde{y}^{(1)}(\tau) - \widetilde{y}^{(1)}(0) = \int_0^{\tau} \widetilde{Y}^{(0)}(\tau) d\tau. \tag{13.20}$$

The following formula can be obtained from (13.19) and (13.20) for $\tau \to \infty$

$$\widetilde{y}^{(1)}(0) = -\int_0^{\infty} \widetilde{Y}^{(0)}(\tau) d\tau.$$

Then from (13.20),

$$\widetilde{y}^{(1)}(\tau) = -\int_{\tau}^{\infty} \widetilde{Y}^{(0)}(\tau)d\tau.$$

From the first equation of (13.18)

$$\overline{y}^{(1)}(0) = -\widetilde{y}^{(1)}(0) = \int_{0}^{\infty} \widetilde{Y}^{(0)}(\tau)d\tau. \qquad (13.21)$$

The expression (13.21) can be treated as the deviation with respect to the slow variable y, accumulated due to the transient in the boundary layer.

Now it is possible to find $\overline{z}^{(1)}(0)$ from the second equation of (13.16) for $t = 0$, and then to find the initial conditions $\widetilde{z}^{(1)}(0) = -\overline{z}^{(1)}(0)$ from the second equation of (13.18).

This terminates the construction of equations of the first approximation.

The equations of next approximations are constructed in just the same manner. The equations of the arbitrary kth approximation are the linear system; its homogeneous part has a form similar to (13.16) and (13.17), and the summands, explicitly depending on time, are calculated with the help of (13.7) applying the already found previous approximations. The procedure of the initial condition determination is similar to the first approximation: the problem definition is completed by the condition $\widetilde{y}^{(k)}(\tau) \to 0$ as $\tau \to \infty$, and then

$$\overline{y}^{(k)}(0) = \int_{0}^{\infty} \widetilde{Y}^{(k-1)}(\tau)d\tau, \quad \widetilde{z}^{(k)}(0) = -\overline{z}^{(k)}(0). \qquad (13.22)$$

In (13.22) the corresponding component of the expansion of function $\widetilde{Y}$ from (13.7) is denoted by $\widetilde{Y}^{(k-1)}$.

Thus, the procedure of the construction of the expansion components (13.2), (13.5), and (13.6) is fully defined.

13.1.2 A. B. Vasil'eva has proved some important properties of this expansion [26].

1. Estimation of the remainder term.

By means of the preceding algorithm N terms of the asymptotic expansion (13.2), (13.5), and (13.6) with respect to variables y and z can be constructed:

$$y_{(N)}(t,\mu) = \sum_{k=0}^{N} \mu^k \left[\overline{y}^{(k)}(t) + \widetilde{y}^{(k)}(\tau)\right],$$

$$z_{(N)}(t,\mu) = \sum_{k=0}^{N} \mu^k \left[\overline{z}^{(k)}(t) + \widetilde{z}^{(k)}(\tau)\right].$$

Then the following evaluation is valid.

$$\begin{aligned} &\|y(t,\mu) - y_{(N)}(t,\mu)\| < C\mu^{N+1}, \\ &\|z(t,\mu) - z_{(N)}(t,\mu)\| < C\mu^{N+1}, \quad 0 \le t \le t', \quad \mu \to 0. \end{aligned} \tag{13.23}$$

Here C is the positive constant, and t' is the value specified by the formulation of the Tikhonov theorem. In this case it follows from (13.23) that for all components of the vectors y and z the difference between exact and approximated solutions will be the value of order μ^{N+1}.

2. The estimation of the boundary functions.

For the summands of the expansion (13.16) inside the boundary layer the following estimation is correct.

$$\begin{aligned} &\|\widetilde{y}^{(k)}(\tau)\| \le ae^{-\lambda\tau}, \\ &\|\widetilde{z}^{(k)}(\tau)\| \le ae^{-\lambda\tau}, \end{aligned} \tag{13.24}$$

when $\tau \ge 0$ and for all $k = 1, \ldots, N$. Here a, λ are positive constants.

The addition to the Tikhonov theorem follows from the estimations (13.22) and (13.24) and the condition $\widetilde{y}^{(0)} \equiv 0$: let the condition of the asymptotic stability of the stationary point of the adjoined system be satisfied according to the equations of linear approximation. Then the difference between the solutions of the initial and degenerate systems according to Tikhonov will be the value of order μ with respect to variable y at the interval $0 \le t \le t'$ and variable z on $0 < t \le t'$.

13.2 Asymptotic expansions of solution for the Stokes problem

Let us consider the problem connected with the vertical fall of a ball in a viscous liquid from Sec. 2.1.2. The equations of the motion can be written in the form similar to (2.7).

The class of the motion "in large" is considered, when characterizing velocity V_* is equal to the velocity of the stationary fall $V_1 = MG/K$, and the characterizing time is $T_* \gg T_1 = M/K$. Normalization of this class leads in Sec. 2.1.2 to the equations:

$$\begin{aligned} &\mu\frac{dv}{dt}1 - v; \qquad v(0) = v_0, \\ &\frac{dx}{dt} = v; \qquad x(0) = x_0, \\ &\mu = \frac{T_1}{T_*} \ll 1. \end{aligned} \tag{13.25}$$

Equations of zeroth approximation can be obtained according to (13.14) and (13.15). Outside the boundary layer

$$\frac{d\overline{x}^{(0)}}{dt} = \overline{v}^{(0)}, \quad 0 = 1 - \overline{v}^{(0)}; \quad \overline{x}^{(0)}(0) = x_0;$$

whereas inside the boundary layer

$$\frac{d\widetilde{v}^{(0)}}{d\tau} = -\widetilde{v}^{(0)}; \quad \widetilde{v}^{(0)}(0) = v_0 - 1.$$

From the preceding equations it follows that:

$$\overline{x}^{(0)} = x_0 + t, \quad \overline{v}^{(0)} = 1; \quad \widetilde{v}^{(0)} = (v_0 - 1)e^{-\tau}. \tag{13.26}$$

According to (13.16), the equations of the first approximation outside the boundary layer are:

$$\frac{d\overline{x}^{(1)}}{dt} = \overline{v}^{(1)}, \quad 0 = -\overline{v}^{(1)}, \quad \overline{x}^{(1)}(0) = v_0 - 1. \tag{13.27}$$

The initial condition in (13.27) is calculated according to (13.21) and (13.26):

$$\overline{x}^{(1)}(0) = \int_0^\infty \widetilde{Y}^{(0)}(\tau)d\tau = \int_0^\infty \widetilde{v}^{(0)}d\tau = \int_0^\infty (v_0 - 1)e^{-\tau}d\tau = v_0 - 1.$$

From (13.27) it follows that $\overline{x}^{(1)} = v_0 - 1$. Therefore, after taking into account (13.26), the partial sum of the two components of the expansion x outside the boundary layer can be written in the form:

$$\overline{x}_{(1)} = x_0 + t + \mu(v_0 - 1). \tag{13.28}$$

Let us compare the obtained approximation with the exact solution for x of the initial system (13.25).

$$\overline{x}_{(1)} = x_0 + t + \mu(v_0 - 1) + \left[(1 - v_0)\mu e^{-t/\mu}\right]. \tag{13.29}$$

The summand in (13.29), included in the square brackets, is the function of the boundary layer. The comparison of (13.28) and (13.29) shows that the slow components of the motion for both expressions coincide. The small addition $\mu(v_0 - 1)$ to the coordinate x and to the law of regular motion $x_0 + t$ appear due to noncoincidence of the speed in the boundary layer with the stationary speed of the fall.

13.3 Asymptotic expansions on the problem of pendulum motion in a medium of high viscosity

Let us write Eq. (12.9) obtained in Sec. 12.1.3 after normalization of the initial equations for the class of slow pendulum motions:

$$\mu\frac{d\omega}{dt} = -\sin\varphi - \omega; \qquad \omega(0) = T_2\Omega_0 \equiv \omega_0,$$
$$\frac{d\varphi}{dt} = \omega, \qquad \varphi(0) = \Phi_0 \equiv \varphi_0, \tag{13.30}$$
$$\mu = \frac{T_1}{T_2} \ll 1.$$

According to (13.14), the equations of zeroth approximation outside the boundary layer have the form:

$$0 = -\sin\overline{\varphi}^{(0)} - \overline{\omega}^{(0)}, \qquad \frac{d\overline{\varphi}^{(0)}}{dt} = \overline{\omega}^{(0)}; \tag{13.31}$$
$$\overline{\varphi}^{(0)}(0) = \varphi_0.$$

Naturally, Eqs. (13.31) coincide with degenerate ones according to the Tikhonov equations (12.13). Excluding in (13.31) the variable $\overline{\omega}^{(0)}$ produces:

$$\frac{d\overline{\varphi}^{(0)}}{dt} = -\sin\overline{\varphi}^{(0)}; \tag{13.32}$$
$$\overline{\varphi}^{(0)}(0) = \varphi_0.$$

The equations of zeroth approximation inside the boundary layer according to (13.15) are:

$$\frac{d\widetilde{\omega}^{(0)}}{d\tau} = -\widetilde{\omega}^{(0)}; \tag{13.33}$$
$$\widetilde{\omega}^{(0)}(0) = \omega_0 + \sin\varphi_0.$$

From (13.33) it follows that

$$\widetilde{\omega}^{(0)} = (\omega_0 + \sin\varphi_0)e^{-\tau}. \tag{13.34}$$

According to (13.16) the equations of the first approximation outside the boundary layer are

$$\frac{d\overline{\omega}^{(0)}}{dt} = -\overline{\omega}^{(1)} - \cos\overline{\varphi}^{(0)}\,\overline{\varphi}^{(1)}, \tag{13.35}$$
$$\frac{d\overline{\varphi}^{(1)}}{dt} = \overline{\omega}^{(1)}.$$

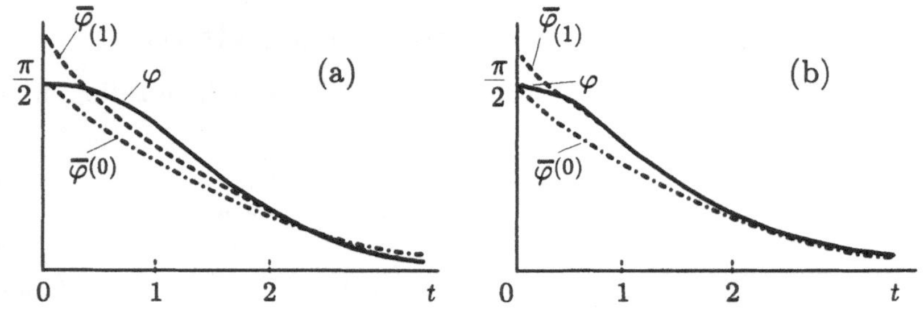

Fig. 18

Substituting the value $\overline{\omega}^{(0)}$ from (13.31) in the first equation and excluding the variable $\overline{\omega}^{(1)}$ in (13.35), yields:

$$\frac{d\overline{\varphi}^{(1)}}{dt} = -\cos\overline{\varphi}^{(0)}\,\overline{\varphi}^{(1)} - \sin\overline{\varphi}^{(0)}\cos\overline{\varphi}^{(0)}. \tag{13.36}$$

The initial value is defined according to (13.21) and (13.34):

$$\overline{\varphi}^{(1)}(0) = \int_0^\infty \widetilde{\omega}^{(0)}d\tau = \omega_0 + \sin\varphi_0. \tag{13.37}$$

Let us construct only the partial sum

$$\overline{\varphi}_{(1)}(t) = \overline{\varphi}^{(0)}(t) + \mu\overline{\varphi}^{(1)}(t) \tag{13.38}$$

and then estimate the accuracy of the approximation for the concrete numeric values of the small parameter.

Figure 18(a) for $\mu = 0.1$, $\varphi(0) = \pi/2$, $\omega(0) = 0$ contains the plots of the dependence of the exact solution for φ of the system (13.30) as a function of time, and dependence of the approximate solutions $\overline{\varphi}^{(0)}$ of the system (13.32) and $\overline{\varphi}_{(1)}$ of the system (13.38), (13.32), (13.36), and (13.37) as a function of time. Figure 18(b) contains the analogous tracing for $\mu = 0.25$ and the same initial conditions. Integration of the considered equation was performed on the computer. The comparison of the plots makes it possible to conclude that for the given problem asymptotic expansions provide satisfactory accuracy of the approximation for the quite large values of the small parameter. Further, it is seen that the second component of the expansion may be used to detect effectively the deviation due to the transient in the boundary layer. This phenomenon becomes essential for the "large" values of the small parameter.

13.4 Decomposition of motions of a railway car in magnetic suspension

Let us consider the problem concerning the plane motion of a car with force suspension, performed by means of controlled electromagnets, installed on the multilinked intermediate ski [31]. The car itself is suspended above the ski by means of soft elastic-damping elements, applied for amortization (Fig. 19). The electromagnets are controlled according to the registration results of the sensors, which detect the deviations of the ski position from the ferromagnetic rail. The application of such a two-step scheme provides more comfort and a more exact following of the path irregularity as opposed to the case when electromagnets are installed directly on the car. This increases the lifting efficiency of the car.

The preceding system has many degrees of freedom, and its analysis, even preliminary, is quite difficult.

It should be noted that appearance in the system of the motion components with strongly differing frequencies, that is, the fast ones due to work of the stiff control system, and slow ones due to soft amortization, is expected.

To evaluate the possibility of motion decomposition, a simplified problem [32] is considered, where the dynamics of the system are simulated by the scheme consisting of two masses, shown in Fig. 20.

The ferromagnetic rail is shown by the hatched line in Fig. 20. Behavior of the electromagnetic suspension is simulated here by the inertialess elastic-damping element I. The car 2 is attached to the ski 1 by means of elastic-damping suspension II.

The equations of motion of the model in Fig. 20 can be written in the

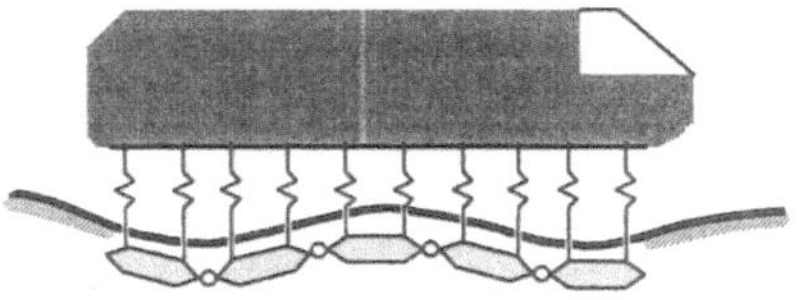

Fig. 19

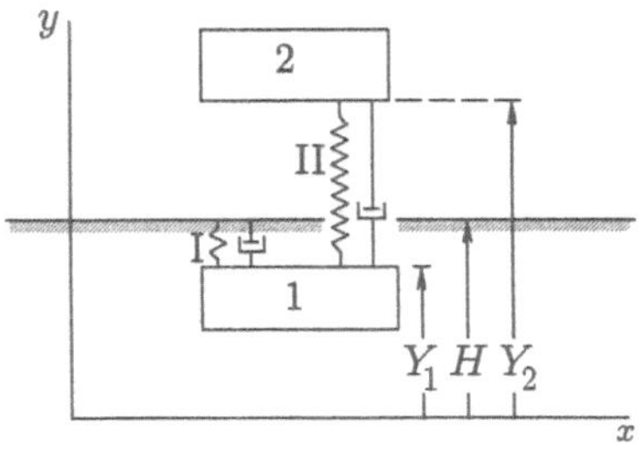

Fig. 20

form:

$$\begin{aligned} M_2\frac{d^2Y_2}{dT^2} &= -P_2 - K_2(Y_2 - Y_1) - R_2\frac{d}{dT}(Y_2 - Y_1), \\ M_1\frac{d^2Y_1}{dT^2} &= -P_1 + K_2(Y_2 - Y_1) + R_2\frac{d}{dT}(Y_2 - Y_1) \\ &\quad -K_1(Y_1 - H) - R_1\frac{d}{dT}(Y_1 - H). \end{aligned} \tag{13.39}$$

Here Y_2 and Y_1 are the vertical coordinates of the car and the ski at the frame of reference, connected with the nominal position of the ski; H is the coordinate defining the actual position of the rail section that interacts with the suspension. The value $H(T/T_{\mathrm{p}})$ changes with time, because the rail has a variable profile due to technological, mounting and other imperfections, and the model moves along this profile. T_{p} is the characterizing time for this process, defined by the typical length of the rail irregularity and the path speed; M_2, P_2; M_1, P_1 are correspondingly the mass and weight of the car and the ski; and K_2, R_2; K_1, R_1 are the coefficients of the stiffness and damping for elastic-damping and magnetic suspensions.

Dividing the equations of the system (13.39) respectively by the K_2 and K_1 leads to:

$$\begin{aligned} T_2^2\frac{dV_2}{dT} &= -\frac{P_2}{K_2} - (Y_2 - Y_1) - 2\zeta_2 T_2(V_2 - V_1), \\ T_1^2\frac{dV_1}{dT} &= -\frac{P_1}{K_1} + \frac{K_2}{K_1}(Y_2 - Y_1) + \frac{R_2}{K_1}(V_2 - V_1) \\ &\quad -(Y_1 - H) - 2\zeta_1 T_1(V_1 - \frac{d}{dT}H), \\ \frac{dY_2}{dT} &= V_2, \qquad \frac{dY_1}{dT} = V_1. \end{aligned} \tag{13.40}$$

Here $T_2^2 = M_2/K_2$, $T_1^2 = M_1/K_1$, $2\zeta_2 T_2 = R_2/K_2$, $2\zeta_1 T_1 = R_1/K_1$.

The system (13.40) should be normalized in accordance with the expressions:

$$\begin{aligned} t &= \frac{T}{T_*}, \qquad y_2 = \frac{Y_2}{Y_{2*}}, \quad v_2 = \frac{V_2}{V_{2*}}, \\ y_1 &= \frac{Y_1}{Y_{1*}}, \quad v_1 = \frac{V_1}{V_{1*}}, \quad h = \frac{H}{H_*}. \end{aligned} \tag{13.41}$$

Let us define the class of motion under consideration. The motion of the car presents the most interest; therefore, it is chosen that $T_* = T_2$. The elastic forces, arising both in the suspension and shock absorber, must have the value of the car weight order: $K_2Y_{2*} = K_1Y_{1*} = P_2$. Under the large diversity of the partial frequencies of the system, the oscillation elements are connected weakly. So the values of the characterizing velocities could be defined, as in Sec. 12.3, by the evaluations for the independent oscillation elements: $V_{2*} = Y_{2*}/T_2$, $V_{1*} = Y_{1*}/T_1$. The typical irregularity of the rail profile is considered as the value of the typical rail shift order: $H_* = Y_{1*}$. The perturbations of the path are assumed to be slow: $T_{\text{P}} \sim T_2$. By virtue of initial assumptions, the relations M_1/M_2, K_2/K_1 are small. For simplicity it is assumed that $M_1/M_2 = K_2/K_1 = \mu \ll 1$. Therefore, $T_1/T_2 = \mu$.

After the normalization (13.41), Eqs (13.40) take the form:

$$\begin{aligned}
\frac{dv_2}{dt} &= -1 - (y_2 - \mu y_1) - 2\zeta_2(v_2 - v_1), \\
\mu\frac{dv_1}{dt} &= -\mu + (y_2 - \mu y_1) + 2\zeta_2(v_2 - v_1) \\
&\quad -(y_1 - h(\lambda t)) - 2\zeta_1\left[v_1 - \mu\frac{d}{dt}h(\lambda t)\right], \\
\frac{dy_2}{dt} &= v_2, \quad \mu\frac{dy_1}{dt} = v_1; \quad \mu = \frac{T_1}{T_2} \ll 1, \quad \lambda = \frac{T_2}{T_{\text{P}}} \sim 1.
\end{aligned} \tag{13.42}$$

The system (13.42) is singularly perturbed. Let us construct asymptotic approximations of its solution.

The equations of zeroth approximation outside the boundary layer are:

$$\begin{aligned}
\frac{d\overline{v}_2^{(0)}}{dt} &= -1 - \overline{y}_2^{(0)} - 2\zeta_2\overline{v}_2^{(0)}, & \frac{d\overline{y}_2^{(0)}}{dt} &= \overline{v}_2^{(0)}, \\
0 &= \overline{y}_2^{(0)} + 2\zeta_2\overline{v}_2^{(0)} - (\overline{y}_1^{(0)} - h), & \overline{v}_1^{(0)} &= 0.
\end{aligned} \tag{13.43}$$

The first two equations of the system (13.43) do not depend on the ski motion and the parameters of the path. They describe free decreasing oscillations of the car in its amortization system. In addition, Gradstein conditions from Sec. 11.3 are satisfied, and Eqs. (13.43) are also valid for the infinite time interval t. Due to asymptotic stability of this oscillations $\overline{y}_2^{(0)}(t)$, $\overline{v}_2^{(0)}(t) \to 0$ as $t \to \infty$.

The last equation (13.43) is the equation of the quasistatic force balance. It defines the shift of the ski $\overline{y}_1^{(0)}$. After the free oscillations with respect to $\overline{y}_2^{(0)}$, $\overline{v}_2^{(0)}$ are damped, we have

$$\overline{y}_1^{(0)} = h(\lambda t). \tag{13.44}$$

The equations of zeroth approximation for (13.42) inside the boundary layer are:

$$\begin{aligned}
\frac{d\widetilde{v}_1^{(0)}}{d\tau} &= y_2(0) + 2\zeta_2(v_2(0) - \widetilde{v}_1^{(0)}) + \left[(h - \widetilde{y}_1^{(0)}) - 2\zeta_1\widetilde{v}_1^{(0)}\right], \\
\frac{d\widetilde{y}_1^{(0)}}{d\tau} &= \widetilde{v}_1^{(0)}.
\end{aligned} \tag{13.45}$$

It is clear from (13.45), that for the approximated analysis of fast motions with the error of μ, the slow variables y_2, v_2, h of the initial equations (13.42) have to be considered as fixed.

Now the influence of the path irregularities on the low frequency motion should be evaluated. Let us form for (13.42) the equations of the approximations of higher orders outside the boundary layer. Only those equations should be written that are necessary for the preceding evaluation:

$$\frac{d\overline{v}_2^{(1)}}{dt} = -2\zeta_2\overline{v}_2^{(1)} - \overline{y}_2^{(1)} + 2\zeta_2\overline{v}_1^{(1)} + \overline{y}_1^{(0)},$$
$$\frac{d\overline{y}_2^{(1)}}{dt} = \overline{v}_2^{(1)}, \ldots, \overline{v}_1^{(1)} = \frac{d\overline{y}_1^{(0)}}{dt}. \tag{13.46}$$

Exclusion of the variables $\overline{v}_1^{(1)}$, $\overline{v}_2^{(1)}$ in (13.46) leads to:

$$\frac{d^2\overline{y}_2^{(1)}}{dt^2} + 2\zeta_2\frac{d\overline{y}_2^{(1)}}{dt} + \overline{y}_2^{(1)} = -\left(\overline{y}_1^{(0)} + 2\zeta_2\frac{d\overline{y}_1^{(0)}}{dt}\right). \tag{13.47}$$

After substitution of the value $\overline{y}_1^{(0)}$ from (13.44) into (13.47), the obtained equation defines the forced component of the car motion (due to path irregularities), which is determined after finishing its own oscillations.

Let us consider the more difficult model of the system, shown in Fig. 19, taking into account the ladder-type structure of the ski, the finite size of the car, and real characteristics of the control system for the electromagnetic suspension. It is assumed that partial frequencies of the ski control system significantly exceed the partial frequencies of the car at the amortization system. Then for the approximate investigation of this model the same methods as for the model of Fig. 20 can be applied.

Remark. Let us point out the important methodical way that must be applied while analyzing the complex model of Fig. 19. In the course of its investigation, there is no need to repeat all calculations performed while working with the simple testing model of Fig. 20. It is possible to write immediately the simplified equations for the special problems that arise while analyzing the system.

Thus, while studying the slow motions of the car in the amortization system, it is necessary to form equations similar to (13.43) which describe the car motion under the action of the main vector and main momentum of the forces generated by all elements of the amortization system. In this case the imperfections of the path and the deviation of the ski from the path are assumed to be zero.

While investigating the dynamics of the suspension control system, equations like (13.45) are written for the ski, "gripped" between the stationary car and the rail. The system has the chain structure and its equations are symmetrical, which makes it possible to choose the control of the suspension [33].

While investigating the forced oscillations of the car, equations similar to (13.46) are written, where the high frequency dynamics of the suspension are not taken into account. The car is perturbed by the shifts of the mounting points of the elements of the amortization system to the ski, which are determined by the kinematics relations analogous to (13.44).

It is important that during the stage of investigation of the problems connected with the dynamics of the model of Fig. 19, it is not necessary to construct full initial equations similar to (13.42). Simplified equations are written for each particular problem. The simplifying mechanical assumptions, accepted during construction of such particular systems of the equations, are formed during the process of accurate asymptotic analysis of the test model of Fig. 20.

The rejection of constructing a full initial equation is extremely valuable for the investigation of complex ladder-type systems. The "complete" equations of such systems can contain hundreds and thousands of summands. The examples of stage investigation of the complex systems are the gyroscopic [34] systems, robots [35], and so on.

CHAPTER V

Decomposition of motion in systems with discontinuous characteristics

14 Definition of a solution in discontinuity points

The existence and uniqueness theorem does not define the solution in points where the right-hand sides of differential equations are discontinuous. Some methods of definition completing the solution in this case were considered in Filippov [36], Utkin [37], and Gerashenko and Gerashenko [38]. In this section the method of definition from Novozhilov [39] is discussed, such that there exists a normalization which carries the problem into the Tikhonov form considered previously.

The following system of differential equations is analyzed.

$$\frac{dx_k}{dt} = f_k(x_1, \dots, x_n, t), \qquad k = 1, \dots, n. \tag{14.1}$$

It is assumed that $x_1, \dots, x_n, t$ have been already normalized and have the values of unity order. Let the function f_j be discontinuous. For simplicity it is supposed that the surface of discontinuity is $x_j = 0$.

If f_j is discontinuous at the surface $G(x_1, \dots, x_n) = 0$, then substituting $u = G(x_1, ...x_n)$ transforms the problem to the preceding case. Function f

$$\lim_{x_j \to +0} f_j = f_j^+, \quad \lim_{x_j \to -0} f_j = f_j^-, \quad f_j^+ \neq f_j^-.$$

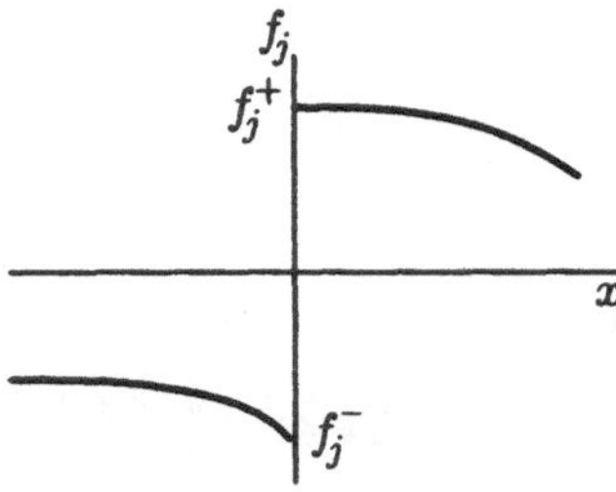

Fig. 21

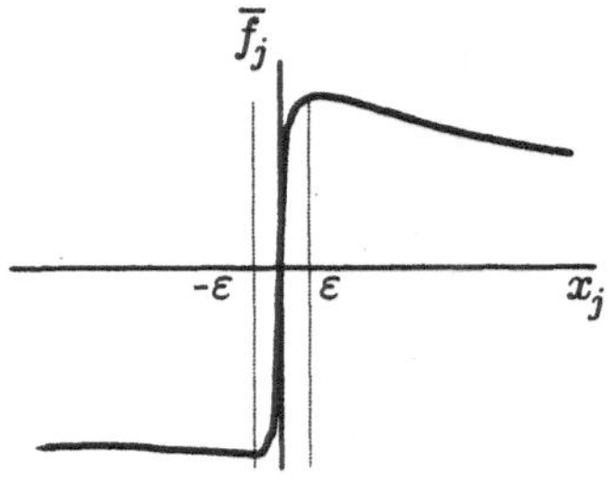

Fig. 22

has different values at different sides of the discontinuity surface (Fig. 21). The functions are smooth with respect to all other variables.

Let us define auxiliary functions $\overline{f}_j$. Functions $\overline{f}_j$ coincide with f_j outside a small vicinity $|x_j| \leq \varepsilon$; inside the vicinity the functions $\overline{f}_j$ connect the values of f_j at the boundaries of the domain $|x_j| \leq \varepsilon$ by an arbitrary sufficiently smooth manner (Fig.22). Then instead of (14.1) the following system may be considered.

$$\begin{aligned} \frac{dx_i}{dt} &= f_i(x_1, \ldots, x_j, \ldots, x_n, t), \\ \frac{dx_j}{dt} &= \overline{f}_j(x_1, \ldots, x_j, \ldots, x_n, t), \quad i \neq j. \end{aligned} \tag{14.2}$$

All functions in the system (14.2) are continuous. Its solution is determined by the existence and uniqueness theorem.

Solutions of system (14.2) as $\varepsilon \to 0$ are proposed in Boltyansky and Pontryagin [40] and Tsypkin [41] as the solutions of the system (14.1).

It should be noted that $|\partial \overline{f} / \partial x_j| \to \infty$ in the domain $|x_j| \leq \varepsilon$ as $\varepsilon \to 0$; therefore, this proposition does not solve the problem of determination of the solution of system (14.2) in the domain $|x_j| \leq \epsilon$.

The next step is the following: we use the appearance of a steep-sided characteristic for introduction of a small parameter and reduction of the system to singularly perturbed form. Now the method of small parameter considered in Sec. 2.2.5 may be used.

The class of motions in the ε-vicinity of the discontinuity point is considered. This class of motions is normalized as:

$$\begin{aligned} z_j &= \frac{x_j}{x_{j*}}, \\ x_{j*} &= \varepsilon. \end{aligned} \tag{14.3}$$

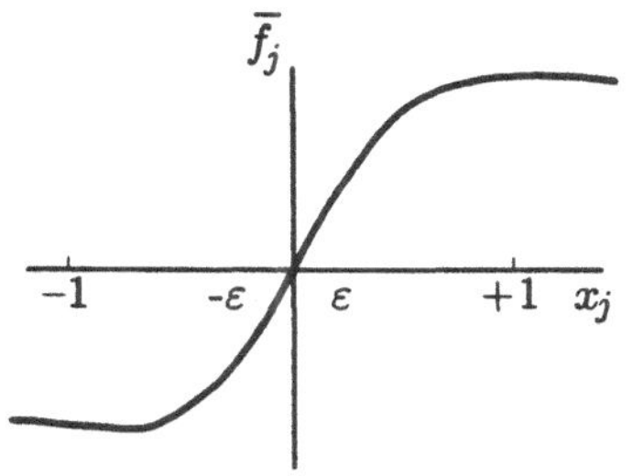

Fig. 23

The dependence of $\overline{f}_j$ on the new variable z_j is shown in Fig. 23. Equations (14.2) take the form:

$$\begin{aligned} \frac{dx_i}{dt} &= f_i(x_1, \ldots, \varepsilon z_j, \ldots, x_n, t), \\ \varepsilon \frac{dz_j}{dt} &= \overline{f}_j(x_1, \ldots, z_j, \ldots, x_n, t), \quad (i \neq j). \end{aligned} \tag{14.4}$$

System (14.4) has a small multiplier by derivative. This system should be investigated by the usual method: the degenerate system for the system (14.4) looks as follows.

$$\begin{aligned} \frac{dx_i}{dt} &= f_i(x_1, \ldots, 0, \ldots, x_n, t), \\ 0 &= \overline{f}_j(x_1, \ldots, z_j, \ldots, x_n, t), \quad (i \neq j). \end{aligned} \tag{14.5}$$

After taking into account the last equation, it is seen that the motion of the system (14.5) lies on the surface of discontinuity $x_j = 0$. Now the system should be tested for satisfying conditions of Tikhonov's theorem. If these conditions are satisfied, then there exists a motion satisfying the system (14.5). Thus, for the system (14.1) on the surface of discontinuity there is realized a fast so-called "sliding mode." In this case the first equation of system (14.5) defines a slow motion on the discontinuity surface, that is. the motion under the constraint $x_j = 0$. The "reaction force" of the constraint is given by the second equation of the system (14.5).

Otherwise, if Tikhonov's theorem conditions are not satisfied, then the system is not retained on the discontinuity surface and moves through it.

15 Examples

15.1 Relay control of angular motion of spacecraft. Sliding mode

The problem of the control of angular motion of a spacecraft [42] is considered. Control moment is provided by two pairs of jets, which work in relay mode. Switch of moment sign is done by changing sign the of control signal, which is a linear combination of angular state and velocity signals.

Delay in the control device, insensitivity zone, and hysteresis of the relay are neglected. Thus, equations of plane motion of the spacecraft have the form:

$$\frac{d\Omega}{dT} = -\gamma \operatorname{sgn} U, \quad \frac{d\Phi}{dT} = \Omega, \quad U = \Phi + T_0 \Omega. \tag{15.1}$$

Here Φ is the angle of spacecraft deviation from a given direction, Ω is the angular velocity, U is the signal of the control system, and $\gamma = \text{const}$ is the absolute value of acceleration. The aim of control is to minimize Φ and Ω. The parameter T_0 is chosen, with which the problem has a solution. Let us investigate the motion.

The system is preliminary normalized as follows.

$$t = \frac{T}{T_*}, \quad \varphi = \frac{\Phi}{\Phi_*}, \quad \omega = \frac{\Omega}{\Omega_*}, \quad u = \frac{U}{U_*}.$$

Here T_* is the characterizing time such that the system deviates on the characteristic angle Φ_* moving with acceleration γ. Then $\Phi_* = \gamma T_*^2$. It is assumed that $\Omega = \Phi_*/T_*$, and the characteristic value of the control signal equals the characteristic angle $U_* = \Phi_*$. Suppose the motion of the system has a large initial deviation. Then $\Phi_* = U_* = 1$, $T_* = 1/\sqrt{\gamma}$, $\Omega_* = 1/T_*$

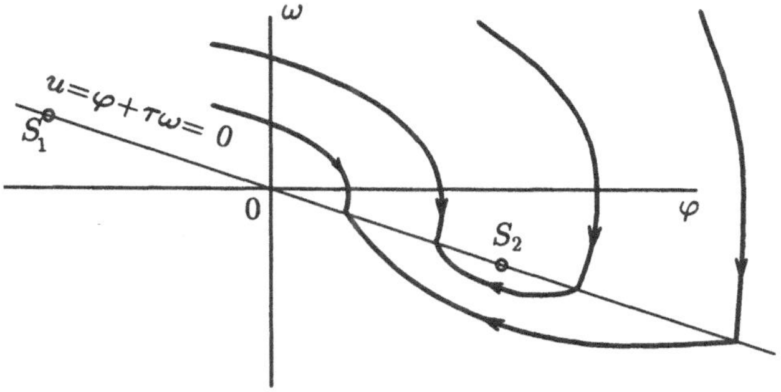

Fig.24

and Eqs. (15.1) take the form:

$$\frac{d\omega}{dt} = -\operatorname{sgn} u, \quad \frac{d\varphi}{dt} = \omega, \quad u = \varphi + \tau_0\omega, \quad \tau_0 = \frac{T_0}{T_*}. \tag{15.2}$$

Behavior of the system (15.2) on the phase plane (φ, ω) is analyzed. By dividing the first equation of the system by the second, it is obtained that the differential equation of phase trajectory has the form:

$$\frac{d\omega}{d\varphi} = -\frac{\operatorname{sgn} u}{\omega}. \tag{15.3}$$

While integrating Eq. (15.3), two cases are considered: $u > 0$, $\operatorname{sgn} u = +1$ and $u < 0$, $\operatorname{sgn} u = -1$. The equations of phase trajectories have the form:

$$\varphi - \varphi(0) = \mp\frac{\omega^2 - \omega^2(0)}{2}. \tag{15.4}$$

Here we take + when $u < 0$ and − when $u > 0$.

Therefore, any phase trajectory is constructed of segments of two families of parabolae (Fig.24). Arrows in Fig. 24 show the direction of system motions along the phase trajectory. Transition from the parabolae of one family to parabolae of the other family occur on the switch line $u = \varphi + \tau_0\omega = 0$. The switch line is divided into intervals of two types. For the interval of the first type the vector of phase velocity is directed toward the switch line at one side, but at the other side the vector of phase velocity is directed from the switch line. In these points the system moves through the switch line. In the interval of the second type the vector of phase velocity is directed toward the switch line at both sides. Probably in this interval the system moves in the sliding mode on the switch line $u = 0$.

Intervals of these two types are separated by points S_1 and S_2 such that the slope of phase trajectory $d\omega/d\varphi = \pm 1/\omega$ (see (15.3)) equals the

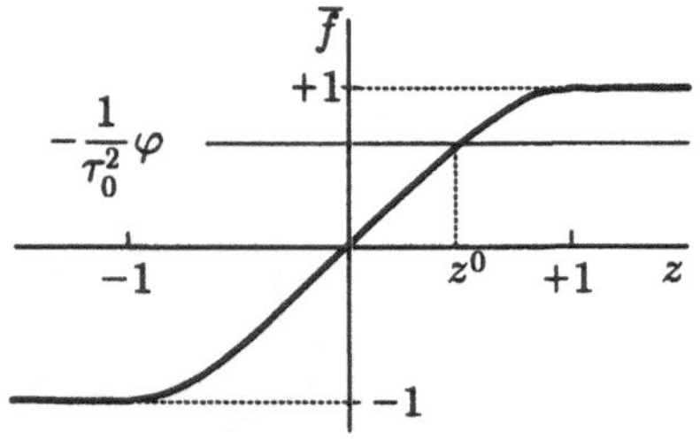

Fig. 25

slope of the switch line $d\omega/d\varphi = -1/\tau_0$. Combining these equations, we get $d\omega/d\varphi = -1/\tau_0$.

Taking into account the switch line equation $\varphi + \tau_0\omega = 0$ for points S_2 and S_1 yields:

$$\varphi = \pm\tau_0^2. \tag{15.5}$$

We now obtain the condition of sliding mode with the help of the method from Sec. 4. In (15.2) the transition from phase variables φ, ω to φ, u is fulfilled by substituting $\omega = (u-\varphi)/\tau_0$. The new set of variables contains u; the right-hand side is discontinuous with respect to this variable. Then the system (15.2) takes the form:

$$\begin{aligned} \frac{du}{dt} &= \frac{1}{\tau_0}(u-\varphi) - \tau_0 \operatorname{sgn} u, \\ \frac{d\varphi}{dt} &= \frac{1}{\tau_0}(u-\varphi). \end{aligned} \tag{15.6}$$

Now a smooth function $\overline{f}(u)$ is substituted for the discontinuous function $\operatorname{sgn} u$; the definition of $\overline{f}(u)$ in the domain $|u| \leq \varepsilon$ is completed in the same way as in (14.2), and the system is normalized in the area $|u| \leq \varepsilon$ by substituting $z = u/\varepsilon$. Then the system (15.6) is transformed by analogy with (14.4) to the form:

$$\begin{aligned} \varepsilon\frac{dz}{dt} &= \frac{1}{\tau_0}(\varepsilon z-\varphi) - \tau_0 \overline{f}(z), \\ \frac{d\varphi}{dt} &= \frac{1}{\tau_0}(\varepsilon z-\varphi). \end{aligned} \tag{15.7}$$

The graph of function $\overline{f}(z)$ is shown in Fig. 25. Equations (15.7) describe fast motion with respect to z near the discontinuity point; the motion with respect to φ is slow.

Now the methods from Sec. 14 are used for obtaining the conditions of appearance of the sliding mode.

The degenerate system corresponding to (15.7) has the form:

$$0 = -\frac{1}{\tau_0}\overline{\varphi} - \tau_0 \overline{f}\,, \qquad \frac{d\overline{\varphi}}{dt} = -\frac{1}{\tau_0}\overline{\varphi}\,. \tag{15.8}$$

The second equation describes the motion along the line of switches, obtained in the preceding from the phase portrait analysis. The variable φ decreases when $\tau_0 > 0$.

The Tikhonov theorem conditions for the system (15.7) should be tested.

Let us form an adjoined system, which describes the system motion with respect to fast variables in fast time. Substituting $\tau = t/\varepsilon$ in (15.7) and assuming that $\varepsilon = 0$ yields:

$$\frac{dz}{d\tau} = -\frac{1}{\tau_0}\varphi - \tau_0 \overline{f}\,(u), \qquad \varphi = \text{const}\,. \tag{15.9}$$

Equilibrium points of (15.9) are determined by the expressions:

$$-\frac{1}{\tau_0}\varphi - \tau_0 \overline{f}\,(z) = 0, \qquad \varphi = \text{const}\,. \tag{15.10}$$

The absolute value of $\overline{f}\,(z)$ is not greater than unity; therefore, Eq. (15.10) has the solution $z = z^0$ only when $|\varphi| \leq \tau_0^2$. This interval is bounded by $\varphi = \pm\tau_0^2$. This condition of existence of the sliding mode coincides with the condition (15.5), which is obtained by geometrical reasoning.

Let us analyze the stability of the equilibrium point z^0. Stability for the equations of the first approximation is investigated by substituting $z = z^0 + \Delta z$ in (15.9). The linear equation with respect to Δz has the form:

$$\frac{d\Delta z}{d\tau} = -\tau_0 k \Delta z, \qquad k = \left[\frac{d\overline{f}}{dz}\right]_{z=z^0}.$$

Taking into account Fig. 25, we get $k > 0$. Thus, for $\tau_0 > 0$ any equilibrium point is asymptotically stable. This stability condition provides the existence of the sliding mode on the switch line if $|\varphi| \leq \tau_0^2$.

15.2 Disc rolling motion with Coulomb friction

A plane disc motion due to the force of a spring is applied in the mass center of the disk C and the force of Coulomb friction F is applied in the point P (Fig. 26).

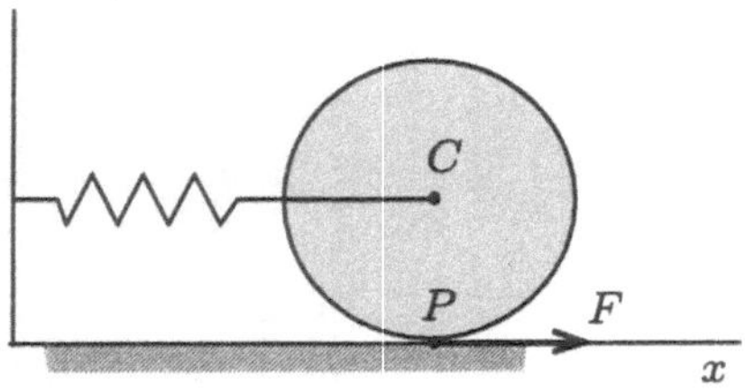

Fig. 26

The equation of motion of the mass center C and the equation of angular momentum with respect to C are used for the description of motion of this system.

$$M\frac{dV}{dT} = F(U) - KX - NV, \quad \frac{dX}{dT} = V,$$
$$I\frac{d\Omega}{dT} = F(U)R, \qquad U = V + \Omega R. \tag{15.11}$$

In addition to the preceding notation the mass and the radius of the disk are denoted by M and R, the coefficient of elasticity and damping are denoted by K and R, X and V are the projections of the radius vector and the velocity vector of the point C to the x-axis, and U is the velocity of the disc sliding in the point P. The dependence of F on U is assumed to have the form:

$$F(U) = -F_0 \operatorname{sgn} U. \tag{15.12}$$

The point P is moving in either the sliding mode or the rolling mode. Equations (15.11) and (15.12) have solutions when the point P is sliding, but for rolling mode the function $F(U)$ is discontinuous at the point $U = 0$ and the equations are senseless. We complete the definition of a solution using the method from Sec. 14.

Let us substitute the set of variables X, V, U by the set X, V, Ω in (15.11); this set contains the variable U, with respect to which the function (15.12) is discontinuous. Excluding from (15.11) the variable Ω, using the last equation of (15.11) we get:

$$M\frac{dV}{dT} = F(U) - KX - NV, \qquad \frac{dX}{dT} = V,$$
$$M\frac{dU}{dT} = (1 + r^2)F(U) - KX - NV, \qquad U = V + \Omega R. \tag{15.13}$$

In (15.13) the notation $r = R/\rho$ is introduced, where ρ is the center radius of the disk inertia. The last equation is not necessary for closing the system but we retain it.

Preliminary normalization of the system (15.3) is fulfilled by:

$$x = \frac{X}{X_*}, \; t = \frac{T}{T_*}, \; v = \frac{V}{V_*}, \; u = \frac{U}{U_*}, \; \omega = \frac{\Omega}{\Omega_*}, \; f = \frac{F}{F_*} .$$

The characterizing time is chosen to be equal to the time constant of the oscillating unit $T_* = T_0$, where $T_0^2 = M/K$. It can be assumed that the sliding velocity is arbitrary and has the order of the point C velocity, and $U_* = V_* = X_*/T_0$. Let the characteristic angle velocity be $\Omega_* = V_*/R$, and $F_* = \max|F| = F_0$. The characteristic displacement X_* is obtained from the condition of the equality of the characteristic values of Coulomb and elastic forces: $KX_* = F_*$. Hence, $X_* = F_0/K$.

The normalized system has the form:

$$\begin{aligned} \frac{dv}{dt} &= f(u) - x - 2\zeta v, & \frac{dx}{dt} &= v , \\ \frac{du}{dt} &= (1+r^2)f(u) - x - 2\zeta v, & u &= v + \lambda\omega . \end{aligned} \tag{15.14}$$

Here $2\zeta = N/KT_0$, $\lambda = R/X_* = RK/F_0$.

The discontinuous function $f(u)$ is substituted with the smooth function $\overline{f}\,(u)$ in (15.14), and its definition is completed in the domain $|u| \le \varepsilon \ll 1$ by analogy with (14.2). In the system obtained in the domain $|u|l\varepsilon$ by substituting $z = u/\varepsilon$ in accordance with (14.3), we obtain

$$\begin{aligned} \frac{dv}{dt} &= \overline{f}\,(z) - x - 2\zeta v, & \frac{dx}{dt} &= v , \\ \varepsilon\frac{dz}{dt} &= (1+r^2)\overline{f}\,(z) - x - 2\zeta v; & \varepsilon z &= v + \lambda\omega . \end{aligned} \tag{15.15}$$

The degenerate system corresponding to (15.15) is:

$$\begin{aligned} \frac{d\overline{v}}{dt} &= \overline{f}\,(\overline{z}) - \overline{x} - 2\zeta\overline{v}, & \frac{d\overline{x}}{dt} &= \overline{v}, \\ 0 &= (1+r^2)\overline{f}\,(\overline{z}) - \overline{x} - 2\zeta\overline{v} . \end{aligned} \tag{15.16}$$

When $\varepsilon = 0$ the last equation of (15.15) has the form:

$$0 = \overline{v} + \lambda\overline{\omega} . \tag{15.17}$$

Equation (15.17) is an equation of a kinematic constraint in the point P, that is, the condition of nonsliding in this point.

From the last equation of (15.16) the magnitude of Coulomb force realizing the constraint is obtained:

$$\overline{f}\,(\overline{z}) = \frac{\overline{x} + 2\zeta\overline{v}}{(1+r^2)} . \tag{15.18}$$

The variable $\overline{f}(\overline{z})$ is excluded from (15.16) using (15.8); then excluding the variable $\overline{v}$ yields:

$$\frac{1+r^2}{r^2}\frac{d^2\overline{x}}{dt^2}+2\zeta\frac{d\overline{x}}{dt}+\overline{x}=0. \tag{15.19}$$

It is the equation of oscillation of the disk, which is rolling without slipping in the point P.

Now the conditions of the Tikhonov theorem are to be tested. The adjoined system for (15.15) has the form:

$$\frac{dz}{d\tau}=(1+r^2)\overline{f}(z)-x-2\zeta v. \tag{15.20}$$

Here slow variables are $x, v = \text{const}$. Equilibrium positions of the system (15.20) with respect to the variable z are determined by the last equation of (15.16) or from Eq. (15.18). Taking into account that the absolute value of the function $\overline{f}(z)$ is not greater than unity, the following condition of existence of equilibrium positions is obtained.

$$\left|\frac{x+2\zeta v}{1+r^2}\right|\leq 1. \tag{15.21}$$

If the definition of the function $\overline{f}(z)$ is completed monotonically (Fig. 25), then asymptotic stability of the points of equilibrium can be easily proved in just the same way as in Sec. 15.1.

The mechanical sense of Eq. (15.21) is obvious: when the system moves in accordance with (15.20), the reaction of the constraint cannot be greater then the maximal value F_0 of Coulomb friction. If the condition (15.21) is not satisfied, then the system moves out of the vicinity $|z| \leq 1$ of the discontinuity point; that is, the system slides in the point P. In this case, the normalization of the system (15.15) is senseless and the system motion is described by the initial system (15.14).

When $\zeta > 0$, Gradstein's additional condition concerning the infinite time interval from Sec. 11.3 is satisfied, since trajectories of the system (15.9) are asymptotically stable with respect to the initial disturbances.

Now consider the case of a conservative limit model. When $\zeta = 0$, the Gradstein condition is not satisfied and proximity of the solutions of the initial system (15.15) and limit system (15.19) is not provided on the infinite time interval. It can be easily proved by comparing the behavior of solutions of both systems. If $\zeta = 0$, then undamped oscillations are the solution of the system (15.19). With $\zeta = 0$ and arbitrarily small ε, the solution of the system (15.15) is damped (tends to zero) as $t \to \infty$. It is readily seen by checking Gurvitz conditions for the linear part of the characteristic of $\overline{f}(u)$ in (15.15). On the other hand, when $\zeta = 0$, the solution of the system (15.19) is an undamped oscillation.

15.3 Relaxation oscillations of the Froude pendulum

A plane physical pendulum oscillating about a rotating shaft (Fig. 27) is considered.

The equations of motion have the form:

$$I\frac{d\Omega}{dT} = -Pl\sin\Phi - M(U), \qquad \frac{d\Phi}{dT} = \Omega, \qquad U = \Omega - \Omega_0. \tag{15.22}$$

Here, in addition to previous notations, $\Omega_0 = const$ is the angular velocity of the shaft, U is the angular velocity of the pendulum with respect to the shaft, and $M(U)$ is the moment of Coulomb friction. The model of friction is refined in comparison with 15.2. Now we take into account the difference between friction of resting M_0 and friction of sliding M_1. These friction characteristics are shown in Fig. 28.

Now preliminary normalization of Eq. (15.22) are implemented. A class of motion for large angles $\Phi_* = 1$ is under consideration. Let us specify $T_* = T_0$, where T_0 is the time constant of small oscillations of the pendulum without friction: $T_0^2 = M/Pl$. It is assumed that $\Omega_* = \Phi_*/T_* = 1/T_0$. We suppose that in the general case in the process of motion, sliding occurs; therefore, it is assumed that $U_* = \Omega_*$. The normalized system for (15.22) is as follows.

$$\begin{aligned} &\frac{d\omega}{dt} = -\sin\varphi - m(u), \\ &\frac{d\varphi}{dt} = \omega, \qquad u = \omega - \omega_0, \quad \omega_0 = T_0\Omega_0. \end{aligned} \tag{15.23}$$

Characteristic $m(u)$ looks like the characteristic in Fig. 28 if M_0 and M_1 are substituted with $m_0 = M_0/Pl$ and $m_1 = M_1/Pl$. It is assumed that $m_0, m_1 < 1$, since in the opposite case the pendulum and the shaft should rotate together.

Let us investigate the discontinuous system (15.23) by the method from Sec. 14. In (15.23), the discontinuous function $m(u)$ is substituted with

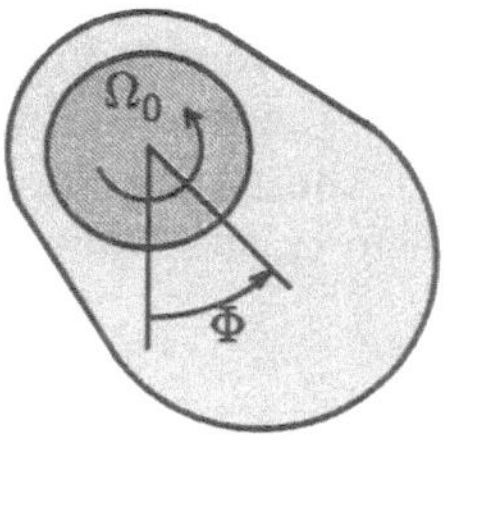

Fig. 27

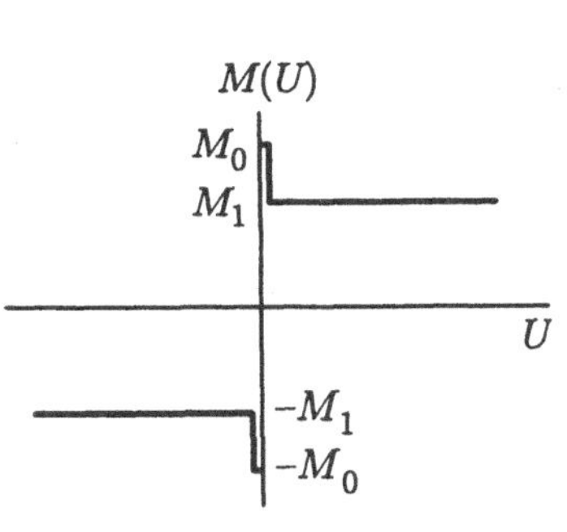

Fig. 28

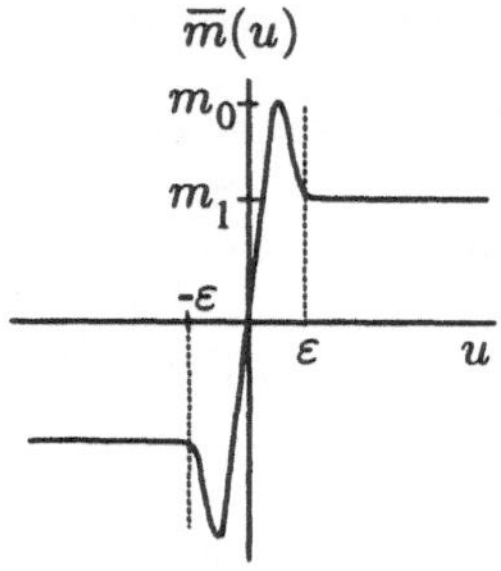

Fig. 29

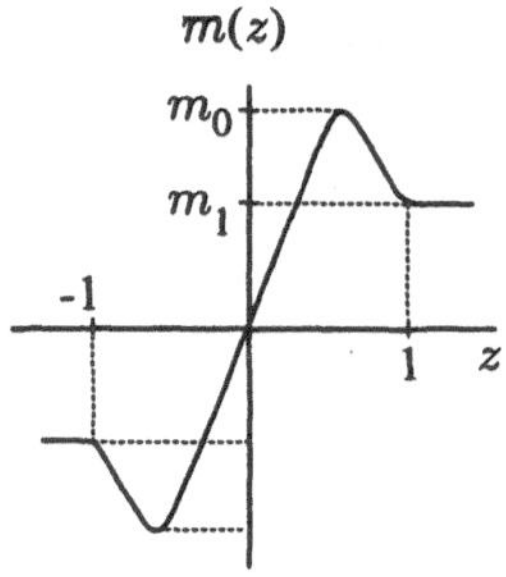

Fig. 30

the function $\overline{m}\,(u)$, presented in Fig. 29. It is supposed that the area of discontinuity is small, $\varepsilon \ll 1$.

The initial set of phase variables φ, ω is substituted with the set φ, u, containing the fast variable u. Supposing the system to be in discontinuity zone $|u| \leq \varepsilon$, then additional normalization $z = u/\varepsilon$ should be implemented. After calculations by analogy with 15.1 and 15.2, the system (15.23) takes the form:

$$\begin{aligned} \varepsilon \frac{dz}{dt} &= -\sin\varphi - \overline{m}\,(z), \\ \frac{d\varphi}{dt} &= \varepsilon z + \omega_0, \qquad \varepsilon z = \omega - \omega_0, \qquad \varepsilon \ll 1. \end{aligned} \tag{15.24}$$

The graph of $\overline{m}\,(z)$ is shown in Fig. 30.

The system (15.24) is singularly perturbed. The degenerate system for it is obtained by supposing $\varepsilon = 0$.

$$\begin{aligned} 0 &= -\sin\overline{\varphi} - \overline{m}\,(\overline{z}), \\ \frac{d\overline{\varphi}}{dt} &= \omega_0, \qquad 0 = \overline{\omega} - \omega_0. \end{aligned} \tag{15.25}$$

Equations (15.25) define the motion of the pendulum together with the shaft, without slipping. Therefore, kinematic constraint is realized.

The conditions of the Tikhonov theorem concerning admissibility of passage to the limit from (15.24) to (15.25) should be checked. Conditions of smoothness 1 and 3 for the right-hand sides of the systems are evidently satisfied. Let us test Condition 2 concerning the existence of an isolated root with respect to z of the equation

$$0 = -\sin\varphi - \overline{m}\,(z), \qquad \varphi = \text{const}. \tag{15.26}$$

Graphically the roots of this equation are defined by the intersection of the graph of function $\overline{m}\,(z)$ with the horizontal line $-\sin\varphi = \text{const}$. It is seen

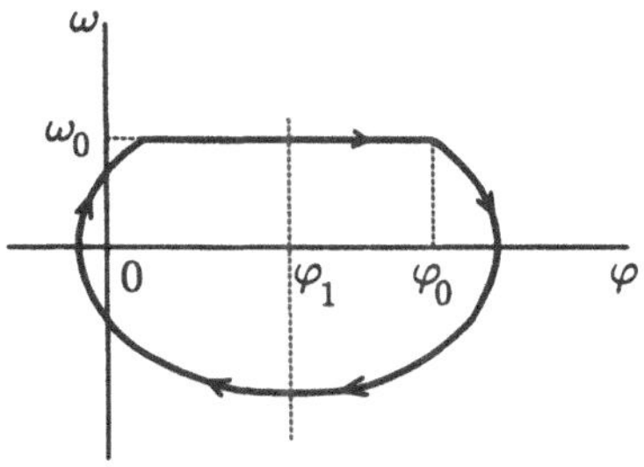

Fig. 31

that when $|\sin\varphi| < m_1$, Eq.(15.26) has a single root with respect to z; when $|\sin\varphi| < m_0$, it has two roots; when $|\sin\varphi| = m_0$, it has a double root; and when $|\sin\varphi| > m_0$, it has no roots. Therefore, the condition

$$|\sin\varphi| \leq m_0 \qquad (15.27)$$

defines the range for the angle $|\varphi| \leq \varphi_0$, where the motion without sliding (15.25) is possible.

For testing the conditions of the Tikhonov theorem, the adjoined system for (15.24) obtained by substituting $\tau = t/\varepsilon$ is used:

$$\frac{dz}{d\tau} = -\sin\varphi - \overline{m}(z), \qquad \varphi = \text{const}. \qquad (15.28)$$

As in the case of Eq. (15.9), it is readily seen that the points of equilibrium for Eq.(15.28) corresponding to ascending parts of characteristics $\overline{m}(z)$ are stable, whereas those corresponding to descending parts are not stable.

Thus, the motion without sliding is realized if the condition (15.27) is satisfied. In the phase plane shown in Fig. 31 this motion takes place in horizontal segment $\omega = \omega_0$, $|\varphi| \leq \varphi_0$. In the point (φ_0, ω_0) the system removes the constraint, because the condition (15.27) of retaining the constraint is violated. Further motion is described by the system (15.23) with $m(u) = -m_1$. After excluding the variable ω from this system it takes the form:

$$\frac{d^2\varphi}{dt^2} + \sin\varphi = m_1. \qquad (15.29)$$

This equation describes free oscillations of the pendulum relative to the shifted position φ_1 defined by the condition $\sin\varphi_1 = m_1$. Since the system (15.29) is conservative, its phase trajectory is symmetric with respect to lines $\varphi = \varphi_1$ and $\omega = 0$. Motion along this trajectory will be continued unless the imaging point again gets into segment $\omega = \omega_0$, $|\varphi| \leq \varphi_0$. After this moment the pendulum will move again without sliding up to point (φ_0, ω_0) and so on.

Periodic oscillations of relaxation type with discontinuity of second derivative are set in the system.

CHAPTER VI

Correctness of limit models

Passage to a limit with respect to a small parameter has already produced degenerate equations describing traditional mechanical models. In Sec. 12.3 a model of absolutely rigid body or, in other words, a model of holonomic constraint was obtained by passage to the limit. A model of kinematic constraint has been also obtained by passage to a limit in Sec. 15.2. In both these examples conditions under which the passage to a limit is possible have been discussed, and we have seen that there exist a lot of cases when these conditions do not hold.

In this section limit models for various particular systems being studied by applied mechanics are obtained and conditions of their correctness are presented.

By a limit model we mean, as previously, a dynamical system defined by equations of zeroth approximation with respect to a small parameter of a problem.

16 Limit model of holonomic constraint (absolutely rigid body)

16.1 Conditions for correctness of the model in statically definable and indefinable cases

During the last centuries applying the models of absolutely rigid body and holonomic constraint has been a holy tradition. Validity of using these models is usually justified by qualitative arguments of passage to a limit as rigidity of a system tends to infinity. In Novozhilov [43, 44] and Chernous'ko [45], limit passage to models of positional constraints has been explained by means of singularly perturbed equations. The starting point of considerations was a system with a given set of generalized coordinates. Such a system was subjected to "hard" effects depending on coordinates. Thus, a preliminary set of holonomic constraints determining generalized coordinates was regarded to be given. However, the problem of correctness of these preliminary constraints remains unsolved.

Let us consider a more general kind of the problem in which a datum is an arbitrary system of the mass point [46]. In this case additional possibilities are available: we may analyze the ambiguity of the choice of generalized coordinates, or construct a limit model of a statically indefinable system.

A system of N mass points is considered. Cartesian coordinates of these points in an absolute frame of reference are denoted by X_i, Y_i, Z_i, $(i=1,\ldots,N)$.

Among all forces acting upon a system there are p "hard" forces of viscous-elastic interaction:

$$F_s = -(K_s + N_s D)\Delta_s; \qquad s = 1,\ldots,p. \tag{16.1}$$

We suppose without significant lack of generality that the forces linearly depend on small relative deformations Δ_s and rate of deformations $D\Delta_s$. The differentiation operator with respect to natural time T is denoted by $D = d/dT$, and the stiffness and damping factors are denoted by K_s and N_s, respectively.

Let L_{ij} be the distance between the ith and jth interacting points and L_{ij}^0 be its value for the nondeformed state of the system. Then

$$\Delta_s = L_{ij} - L_{ij}^0 = \sqrt{(X_i - X_j)^2 + (Y_i - Y_j)^2 + (Z_i - Z_j)^2} - L_{ij}^0. \tag{16.2}$$

If among all interacting points there are exterior ones with respect to the

A Passage to a Limit Model

system, then (16.2) includes their coordinates X_j, Y_j, Z_j represented as explicit functions of time.

The following notation for point coordinates is introduced: X_k ($k = 1, \ldots, n = 3N$). Then in (16.2)

$$\Delta_s = \Delta_s(X_1, \ldots, X_n, T); \qquad s = 1, \ldots, p. \tag{16.3}$$

The relationship between p and n leads to the distinction between statically definable (if $p \leq n$) and statically indefinable (if $p > n$) cases.

Let r be the number of independent functions in (16.3). Then the rank of the matrix $[\partial\Delta_s/\partial X_k]$ is equal to r. For definiteness assume that the submatrix with $s, k = 1, \ldots, r$ is nonsingular.

Consider (16.1) under the assumption that each $K_s \to \infty$. Therefore, the forces in the mechanical systems are finite magnitudes, and we may expect that $K_s \to \infty$ as $\Delta_s \to 0$ and the system will be transformed into its nondeformable absolutely rigid analogue.

The limiting equalities for r independent functions Δ_s from (16.3) have the form:

$$\Delta_s(X_1, \ldots, X_n, T) \equiv 0, \qquad s = 1, \ldots, r. \tag{16.4}$$

As, by assumption, $|\partial\Delta_s/\partial X_k| \neq 0$; $s, k = 1, \ldots, r$, then Eq. (16.4) for $X_1, \ldots, X_r$ are solvable:

$$\begin{aligned} X_1^0 &= X_1^0(X_{r+1}, \ldots, X_n, T), \\ &\ldots \\ X_r^0 &= X_r^0(X_{r+1}, \ldots, X_n, T). \end{aligned} \tag{16.5}$$

Independently defined in (16.5) variables $X_{r+1}, \ldots, X_n$ may be regarded as generalized coordinates of the limit system as $K_s \to \infty$. Dependent coordinates $X_k = X_k^0$ ($k = 1, \ldots, r$) are determined in the limit system by coordinates of exterior points with respect to the system and generalized coordinates.

If the rank of the matrix $[\partial\Delta_s/\partial X_k]$, ($s = 1, \ldots, p$; $k = 1, \ldots, n$) satisfies the preceding condition, then there may exist a few nonsingular submatrices of such dimension. This nonuniqueness leads to nonuniqueness of choice of generalized coordinates of the system.

The equations of the mass-points system motion in Newton form using matrix notation are:

$$M_{(n,n)} \frac{d^2}{dT^2} X_{(n,1)} = A_{(n,p)} F_{(p,1)} + G_{(n,1)}. \tag{16.6}$$

In this equation $X_{(n,1)}$ is the matrix column of variables, $F_{(p,1)}$ is the matrix column of forces (16.1), $M_{(n,n)}$ is the diagonal matrix of point masses, $A_{(n,p)}$ is the matrix defining reprojection of forces (16.1) onto the coordinate axes, and $G_{(n,1)}$ is the column of all other forces acting on the system. Lower indices stand for the numbers of rows and columns.

Let us transform Eqs. (16.1), (16.3), and (16.6) into the singularly perturbed normalized form. As a first step of transformation we replace the given set of variables $X_1, \ldots, X_n$ by the set $\Delta X_1, \ldots, \Delta X_r; X_{r+1}, \ldots, X_n$. In this notation $X_{r+1}, \ldots, X_n$ are slow variables which as $K_s \to \infty$ tend to generalized coordinates of the nondeformable system, and fast variables $\Delta X_1, \ldots, \Delta X_r$ are deviations of dependent ones from their values (16.5) for the nondeformable system:

$$X_k = X_k^0 + \Delta X_k, \qquad k = 1, \ldots, r. \tag{16.7}$$

The forces (16.1) should be expressed by means of new variables. The result of substituting (16.7) into (16.3) and expanding the right-hand sides of obtained equalities up to linear summands with respect to ΔX_k in matrix notation is:

$$\Delta_{(p,1)} = B_{(p,r)} \Delta X_{(r,1)}, \tag{16.8}$$

where $\Delta_{(p,1)}$ and $\Delta X_{(r,1)}$ are the matrix columns of corresponding variables, and $B_{(p,r)} \equiv [\partial \Delta_s / \partial X_k]$, $(s = 1, \ldots, p;\ k = 1, \ldots, r)$. Elements of the matrix $B_{(p,r)}$ depend, generally speaking, on the variables $X_1^0, \ldots, X_r^0$ and $X_{r+1}, \ldots, X_n$. Taking into account (16.8), Eqs. (16.1) may be transformed into

$$F_{(p,1)} = -(K_{(p,p)} + DN_{(p,p)}) B_{(p,r)} \Delta X_{(r,1)}, \tag{16.9}$$

where $K_{(p,p)}, N_{(p,p)}$ are the diagonal matrices of the stiffness and damping coefficients.

Separating the matrix equation into ones of dimensions r and $n - r$, and substituting (16.7) and (16.9) into them leads to:

$$\begin{aligned} M_{(r,r)} \frac{d^2}{dT^2} \Delta X_{(r,1)} &= -A_{(r,p)} (K_{(p,p)} + DN_{(p,p)}) B_{(p,r)} \Delta X_{(r,1)} \\ &\quad + G_{(r,1)} - M_{(r,r)} \frac{d^2}{dT^2} X^0_{(r,1)}, \\ M_{(n-r,n-r)} \frac{d^2}{dT^2} X_{(n-r,1)} & \\ &= -A_{(n-r,p)} (K_{(p,p)} + DN_{(p,p)}) B_{(p,r)} \Delta X_{(r,1)} \\ &\quad + G_{(n-r,1)}. \end{aligned} \tag{16.10}$$

Eliminate in (16.10) the summand containing $d^2 X^0_{(r,1)}/dT^2$, then twice differentiate Eqs. (16.4) and write down the result in a matrix notation

$$B_{(r,r)} \frac{d^2}{dT^2} X^0_{(r,1)} + B_{(r,n-r)} \frac{d^2}{dT^2} X_{(n-r,1)} + \Phi_{(r,1)} = 0, \tag{16.11}$$

where

$$[B_{(r,r)} \vdots B_{(r,n-r)}] = B_{(r,n-r)} \equiv [\partial \Delta_s / \partial X_k], \quad (s = 1, \ldots, r;\ k = 1, \ldots, n),$$

and $\Phi_{(r,1)}$ is the column of other members. From (16.11) we find:

$$\frac{d^2}{dT^2}X^0_{(r,1)} = -B^{-1}_{(r,n-r)}\left[B_{(r,n-r)}\frac{d^2}{dT^2}X_{(n-r,1)} + \Phi_{(r,1)}\right]$$

and substitute it into (16.10). Introduce into (16.10) the phase variables

$$\begin{aligned} \frac{d}{dT}\Delta X_{(r,1)} &= U_{(r,1)}, \\ \frac{d}{dT}X_{(n-r,1)} &= V_{(n-r,1)}. \end{aligned} \tag{16.12}$$

Then (16.10) will be transformed into

$$\begin{aligned} & M_{(r,r)}\frac{d}{dT}U_{(r,1)} \\ &= -A_{(r,p)}[K_{(p,p)}B_{(p,r)}\Delta X_{(r,1)} + N_{(p,p)}B_{(p,r)}U_{(r,1)}] \\ &\quad +G_{(r,1)} + M_{(r,r)}B^{-1}_{(r,r)}\left[B_{(r,n-r)}\frac{d}{dT}V_{(n-r,1)} + \Phi_{(r,1)}\right], \\ & M_{(n-r,n-r)}\frac{d}{dT}V_{(n-r,1)} \\ &= -A_{(n-r,p)}[K_{(p,p)}B_{(p,r)}\Delta X_{(r,1)} + N_{(p,p)}B_{(p,r)}U_{(r,1)}] \\ &\quad +G_{(n-r,1)}. \end{aligned} \tag{16.13}$$

The system of equations (16.12) and (16.13) with respect to the variables $\Delta X_1, \ldots, \Delta X_r$; $X_{r+1}, \ldots, X_n$; $U_1, \ldots, U_r$; and $V_{r+1}, \ldots, V_n$ is closed. These equations are valid in both statically definable and indefinable cases.

Let us normalize the equations obtained, adding to them Eqs. (16.2) unnecessary for the closure of the system:

$$\begin{aligned} t &= \frac{T}{T_*}, \quad x_k = \frac{X_k}{X_*}, \quad \delta x_k = \frac{\Delta X_k}{\Delta X_*}, \\ v_k &= \frac{V_k}{V_*}, \quad u_k = \frac{U_k}{U_*}, \quad m_k = \frac{M_k}{M_*}, \quad k_s = \frac{K_s}{K_*}, \\ n_s &= \frac{N_s}{N_*}, \quad a_{ij} = \frac{A_{ij}}{A_*}, \quad b_{ij} = \frac{B_{ij}}{B_*}, \\ g_k &= \frac{G_k}{G_*}, \quad l_{ij} = \frac{L_{ij}}{L_*}, \quad \delta_s = \frac{\Delta_s}{\Delta_*}. \end{aligned}$$

In order to choose the characterizing values of the quantities $T_*, \ldots, L_*$ simplified estimations similar to ones made in Sec. 12.3 should be performed. It is assumed that damping of fast oscillations determined by the elastic forces (16.1) is small. Then the characterizing value of the fast oscillation time is $T_1 = \sqrt{M_*/K_*}$, and $N_* = 2\zeta_1 T_1 K_*$, where $\zeta_1 < 1$ is the dimensionless damping coefficient. It is assumed, further, that the characterizing values of the elastic forces F_s are the same as that of other forces G_k: $K_*\Delta X_* = G_*$. Let T_2 be a characterizing time of slow components of motion determined by the forces G_k. Select these components and set up $T_* = T_2$. While normalizing the kinematical equations, we assume that $U_* = \Delta X_*/T_1$, $V_* = X_*/T_2$. Normalizing the geometric and trigonometric relations is carried out under the assumption that $X_* = L_*$, $A_* = 1$, $B_* = 1$, $\Delta_* = \Delta X_*$.

Having performed this normalizing, we divide each equation by a factor of its dimension. The result is

$$\begin{aligned}
&\mu m_{(r,r)} \frac{d}{dt} u_{(r,1)} \\
&\quad = -a_{(r,p)}\left[k_{(p,p)} b_{(p,r)} \delta x_{(r,1)} + 2\zeta_1 n_{(p,p)} b_{(p,r)} u_{(r,1)}\right] \\
&\qquad + g_{(r,1)} + m_{(r,r)} b^{-1}_{(r,r)} \left[b_{(r,n-r)} \frac{d}{dt} v_{(n-r,1)} + \varphi_{(r,1)}\right], \\
&m_{(n-r,n-r)} \frac{d}{dt} v_{(n-r,1)} \\
&\quad = -a_{(n-r,p)}\left[k_{(p,p)} b_{(p,r)} \delta x_{(r,1)} + 2\zeta_1 n_{(p,p)} b_{(p,r)} u_{(r,1)}\right] \\
&\qquad + g_{(n-r,1)}, \\
&\mu \frac{d}{dt} \delta x_{(r,1)} = u_{(r,1)}, \quad \frac{d}{dt} x_{(n-r,1)} = v_{(n-r,1)}, \quad \varepsilon \delta_s = l_{ij} - l^0_{ij},
\end{aligned} \tag{16.14}$$

where $\mu = T_1/T_2 \ll 1$, $\varepsilon = \Delta_*/X_* \ll 1$.

The system (16.14) is singularly perturbed. The last equation contains the small parameter ε, which is given, generally speaking, independently of μ. Values of the parameters μ and ε may be compared, if the functions G_k are defined. If, for example, these forces depend linearly on X_k with the small characterizing value of stiffness coefficient K_{2*}, then $T_2 = \sqrt{M_*/K_{2*}}$. By condition $K_*\Delta_* = G_* = K_{2*}X_*$, discussed previously, we obtain $\varepsilon = \Delta_*/X_* = K_{2*}/K_* = \mu^2$.

The system (16.14) may be transformed into Tikhonov form (see Sec. 11), which is solved with respect to the column of derivatives $\mu du_{r,1}/dt$, $dv_{n-r,1}/dt$, $\mu d\delta x_{r,1}/dt$, $dx_{n-r,1}/dt$. The necessary conditions for the transformation are the following: the matrices $m_{(r,r)}$, $m_{(n-r,n-r)}$ must be nonsingular (this always holds), and the inequality

$$|b_{(r,r)}| \neq 0, \tag{16.15}$$

previously discussed must hold. In the following, examples are given where (16.15) does not hold. Assuming that (16.15) is valid, we omit calculations concerning the transformation of (16.4) into Tikhonov form.

The degenerate equations equivalent to (16.14) under the assumption that $\mu, \varepsilon = 0$ have the form:

$$\begin{aligned}
0 &= -a_{(r,p)}[k_{(p,p)}b_{(p,r)}\,\overline{\delta}\,x_{(r,1)} + 2\zeta_1 n_{(p,p)}b_{(p,r)}\,\overline{u}_{(r,1)}] + g_{(r,1)} \\
&\quad + m_{(r,r)}b_{(r,r)}^{-1}\left[b_{(r,n-r)}\frac{d}{dt}\overline{v}_{(n-r,1)} + \varphi_{(r,1)}\right], \\
m_{(n-r,n-r)}&\frac{d}{dt}\overline{v}_{(n-r,1)} \qquad\qquad (16.16)\\
&= -a_{(n-r,p)}[k_{(p,p)}b_{(p,r)}\,\overline{\delta}\,x_{(r,1)} + 2\zeta_1 n_{(p,p)}b_{(p,r)}u_{(r,1)}] + g_{(n-r,1)}, \\
0 &= \overline{u}_{(r,1)}, \qquad \frac{d}{dt}\overline{x}_{(n-r,1)} = \overline{v}_{(n-r,1)}, \qquad 0 = \overline{l}_{ij} - l_{ij}^0.
\end{aligned}$$

The last equation in (16.16) is the equation of holonomic constraints imposed on the system. We may discover $\overline{\delta}\,x_{(r,1)}$ from the first equation. Due to this it is possible to find from the normalized equations (16.9) the forces $f_{(p,1)}$ as functions of inertial forces and forces acting on the systems. In (16.16) the forces $f_{(p,1)}$ are reactions of imposed constraints. Substitution of $\overline{\delta}\,x_{(r,1)}$ into the second equation of (16.16) produces equations describing the motion of the system with constraints, and these equations are expressed in generalized coordinates $\overline{x}_{r+1}, \ldots, \overline{x}_n$.

Now the conditions of the Tikhonov theorem of validity of the limit passage from (16.14) to (16.16) should be verified. Let us assume that the right-hand parts of Eqs. (16.14) satisfy the conditions of smoothness mentioned in the theorem. The equations adjoined to (16.14) should be composed. In order to do this we exclude $dv_{(n-r,1)}/dt$ from the first equation of (16.14) by use of the second one, and introduce the fast dimensionless time $\tau = t/\mu$. The slow variables $x_{(n-r,1)}, v_{(n-r,1)}$ are considered as parameters. The adjoined equations describe fast partial motions with respect to $u_{(r,1)}, \delta_{(r,1)}$ of a mechanical system subjected to actions of constant forces as well as linear elastic and friction forces. Isolated stationary points of the adjoint system exist if the matrix

$$[a_{(r,p)} + m_{(r,r)}b_{(r,r)}^{-1}b_{(r,n-r)}m_{(n-r,n-r)}^{-1}a_{(n-r,p)}]k_{(p,p)}b_{(p,r)} \qquad (16.17)$$

is nonsingular.

Asymptotic stability of stationary points may be verified by means of Chetaev's conditions (see, e.g., Magnus [47]).

If all demands previously formulated are satisfied, then the solutions of the singular system (16.16) and that of the complete one (16.14) are close to each other with the error of the order of mu in a finite time interval. The additional Gradstein condition of t-uniform asymptotic stability of degenerate solutions with respect to initial perturbations guarantees the validity of limit passage in an infinite time interval. Examples where this additional demand is not satisfied are given in Sec. 12.3. That is why one must be careful while using limit models of absolutely rigid body and holonomic constraints for solving conservative problems in an infinite time interval.

16.2 Examples

16.2.1 Consider an absolutely rigid model of an ellipsograph (Fig. 32(a)) and its analogue with finite rigidity of a weightless rod AB and direction slides a and B. The slides possess some masses. The model is shown in Fig. 32(b).

Here $n = 4$, $p = 3$. The following notation is used for the slides of Cartesian coordinates: $X_1 = X_A$, $X_2 = Y_A$, $X_3 = X_B$, $X_4 = Y_B$. The deformations of elastic elements pointed out in Fig. 32(b) are denoted by $\Delta_1, \Delta_2, \Delta_3$. The equalities (16.2) for the model have the form:

$$\begin{aligned} \Delta_1 &= X_2, \\ \Delta_2 &= X_3 - a, \\ \Delta_3 &= \sqrt{(X_3 - X_1)^2 + (X_4 - X_2)^2} - L^0. \end{aligned}$$

The following computations should be done:

$$B_{(3,4)} = \left[\frac{\partial \Delta_s}{\partial X_k}\right] = \begin{pmatrix} 0 & 1 & 0 & 0 \\ 0 & 0 & 1 & 0 \\ -\cos\varphi & -\sin\varphi & \cos\varphi & \sin\varphi \end{pmatrix}, \qquad (16.18)$$

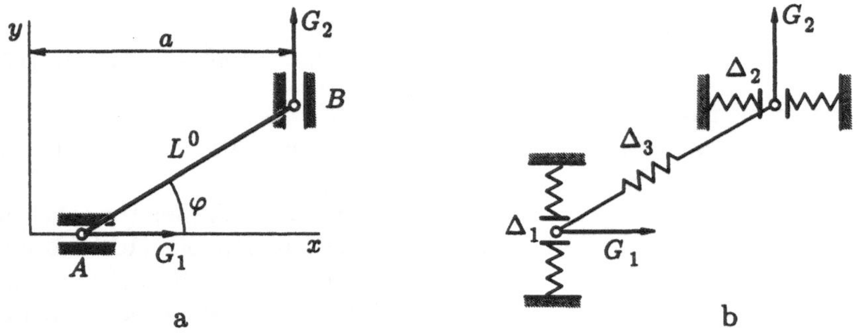

Fig. 32

$$A_{(3,4)} = \begin{pmatrix} 0 & 0 & \cos\varphi \\ 1 & 0 & \sin\varphi \\ 0 & 1 & -\cos\varphi \\ 0 & 0 & -\sin\varphi \end{pmatrix}.$$

It is assumed that the stiffness of all elastic elements is the same. Then

$$K_{(p,p)} = K \begin{bmatrix} 1 & 0 & 0 & 0 \\ 0 & 1 & 0 & 0 \\ 0 & 0 & 1 & 0 \\ 0 & 0 & 0 & 1 \end{bmatrix}.$$

Let us compose $B_{(r,r)}$ of the first three columns of the matrix (16.18). In this case X_4 is the generalized coordinate of the limit model. Condition (16.15) implies the inequality $\cos\varphi \neq 0$. If $\varphi = \pi/2$, then, as seen in Fig. 32(a), motion along the axis X_4 is impossible, and passage to the limit model is not valid.

Then let us compose the matrix $B_{(r,r)}$ of the latter three columns (16.18) which corresponds to the generalized coordinate X_1 in the limit model. Condition (16.15) implies $\sin\varphi \neq 0$. If $\varphi = 0$, then the motion along the axis X_1 is impossible.

Validity of other conditions for passage to the limit, that is, the nonsingularity of matrix (16.17) and the stability of stationary points, is evident. Additional restrictions do not arise.

16.2.2 Let us consider the absolutely rigid model (Fig. 33(a)) and its analogue with finite stiffness (Fig. 33(b)).

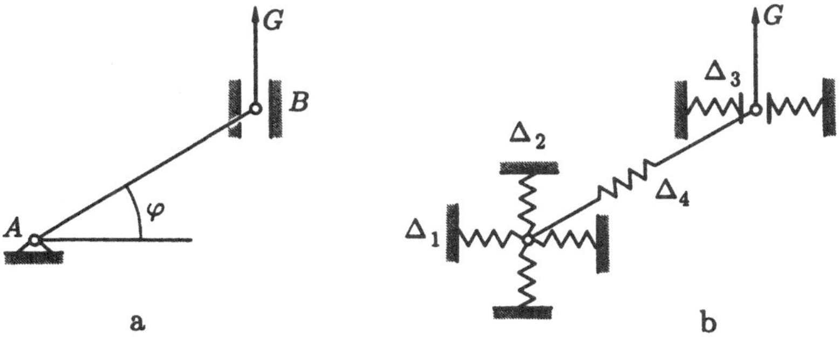

Fig. 33

The case of $p = n$ is considered. The notation for coordinates is similar to the notation in Example 16.2.1. Deformations of elastic elements are pointed out in Fig. 33(b). By analogy with (16.18), let us calculate

$$B_{(4,4)} = \left[\frac{\partial \Delta_s}{\partial X_k}\right] = \begin{pmatrix} 1 & 0 & 0 & 0 \\ 0 & 1 & 0 & 0 \\ 0 & 0 & 1 & 0 \\ -\cos\varphi & -\sin\varphi & \cos\varphi & \sin\varphi \end{pmatrix}.$$

The matrix $B_{(4,4)}$ is nonsingular. The condition (16.15) holds, if $\sin\varphi \neq 0$. Similarly to Example 16.2.1, other conditions for validity of passage to a limit do not generate additional restrictions. Therefore, if $\sin\varphi \neq 0$, the limit passage to a stationary rigid construction represented in Fig. 33(a) is possible. It is seen in Fig. 33(a) that the limit model is not correct if $\varphi = 0$: the geometric constraint imposed on the slide is degenerate and it cannot restrict the side motion.

16.2.3 Let us add to the construction represented in Fig. 33(b) one more rigid-elastic interaction along the axis Y_B. Then in the problem obtained, $n = 4$, $p = 5$, and the limit system becomes statically indefinable. Let $\Delta_1 = X_A$, $\Delta_2 = Y_A$, $\Delta_3 = X_B$, $\Delta_4 = Y_B$, and Δ_5 be the deformation of the rod AB. For simplicity it is assumed that all elastic elements possess the same stiffness. Then $K_{(p,p)} = KE_{(p,p)}$. Let us calculate

$$A_{(4,5)} = \begin{pmatrix} 0 & 0 & 0 & 0 & \cos\varphi \\ 0 & 1 & 0 & 0 & \sin\varphi \\ 0 & 0 & 1 & 0 & -\cos\varphi \\ 0 & 0 & 0 & 1 & -\sin\varphi \end{pmatrix},$$

$$B_{(5,4)} = \begin{pmatrix} 1 & 0 & 0 & 0 \\ 0 & 1 & 0 & 0 \\ 0 & 0 & 1 & 0 \\ 0 & 0 & 0 & 1 \\ -\cos\varphi & -\sin\varphi & \cos\varphi & \sin\varphi \end{pmatrix}.$$

Having verified (16.15) and (16.17), we can see that the conditions for validity of passage to the model of holonomic constraints in this statically indefinable case hold for each φ.

17 Limit model of kinematic constraints

17.1 Conditions of model correctness in kinematically definable and indefinable cases

Kinematic constraints, including nonholonomic ones, occurring in mechanical problems, are specified by prohibition of relative displacements along the tangent to contiguous surfaces. In many works on dynamics of systems with nonholonomic constraints, these constraints are assumed to be primarily imposed on the system. This implies that it is possible to calculate the reaction of constraints and to discover conditions under which they do not exceed limit values of the forces in contact points. The problem can be posed otherwise: to set the forces of this or that nature in contact points and to find conditions under which relative sliding caused by these forces would be impossible. Thus, in Neymark and Fufaev [48] an example is analyzed in which a nonholonomic constraint for the Chaplygin carriage problem is realized by means of viscous friction with an infinite proportionality coefficient. A similar method of realization of nonholonomic constraints was considered in the general statement of the problem in Karapetyan [49] and Kozlov [50].

Coulomb friction was used in Novozhilov [39] and Kalinin and Novozhilov [51] as a force providing the realization of nonholonomic constraints. The problem was reduced to a singularly perturbed one. Now this problem is discussed in a more general statement.

Let us consider a mechanical system on which holonomic bilateral constraints are imposed. Let the system have n degrees of freedom and p kinematic pairs with one relative degree of freedom, in which sliding is allowed by the constraints. (Assumption that the pairs possess only one degree of sliding is not a significant restriction of problem generality. Similar pairs with sliding along two relative degrees of freedom are considered in Kalinin and Novozhilov [51]). It is supposed that in the pairs there is Coulomb friction.

By analogy with statics the cases $p \leq n$ and $p > n$ can be named kinematically definable and kinematically indefinable, respectively.

The equations of system motion in a matrix form are written as:

$$\begin{aligned} \frac{d}{dt}v_{(n,1)} &= a_{(n,p)}f_{(p,1)} + g_{(n,1)}, \\ \frac{d}{dt}x_{(n,1)} &= v_{(n,1)}. \end{aligned} \tag{17.1}$$

The equations are reduced to Cauchy form as required by the Tikhonov theorem. While dealing with concrete systems this auxiliary operation can be omitted; it is sufficient to verify nonsingularity of the inertia coefficient matrix. In (17.1), the variables are assumed to be preliminary normalized; generalized coordinates are denoted by x_k ($k = 1, \dots, n$), and v_k are the velocities of their change; f_s ($s = 1, \dots, p$) are the components of the matrix column of Coulomb friction forces depending on u_s, which are relative velocities in sth pairs; g_k are the other normalized forces acting on the system. Dependence of Coulomb friction on normal reactions is not taken into account.

The following relations are supposed to be a characteristic of friction.

$$f_s(u_s) = -\kappa_s \operatorname{sgn} u_s,$$
$$u_s \neq 0, \qquad \kappa_s = \text{const}, \qquad (s = 1, \dots, p). \tag{17.2}$$

The quantities u_s are related to v_k by the linear kinematic equation:

$$u_{(p,1)} = b_{(p,n)} v_{(n,1)} + c_{(p,1)}. \tag{17.3}$$

Linearity of Eq. (17.3) is the consequence of linearity of the velocitiy summation theorem. Matrices $a_{(n,p)}, b_{(p,n)}, c_{(p,1)}$ in (17.1) and (17.3) depend, generally speaking, on x_k, t, and matrix $g_{(n,1)}$ depends on x_k, v_k, t.

To analyze the system (17.1) – (17.3) with discontinuous functions, the technique introduced in Sec. 14 is to be used. Now we pass in the system to a new set of phase variables. The variables u_s ($s = 1, \dots, r$) of a maximal linearly independent set with respect to which the functions (17.2) are discontinuous, are considered to be fast variables. Here r is the rank of matrix $b_{(p,n)}$ in (17.3). For definiteness it is assumed that $|b_{(r,r)}| \neq 0$. The velocity v_k ($k = r+1, \dots, n$) and positional variable x_k ($k = 1, \dots, n$) are considered to be slow variables.

Decomposition of system variables into subsets of fast and slow ones may be performed nonuniquely. Only dimensions of vectors of fast and slow variables are fixed. Therefore, in this section fast variables could be introduced in another way, that is, by analogy with Sec. 16. However, taking into account the linearity of expressions (17.3) and the tradition of Sec. 14, another procedure is accepted here.

Using the method of Sec. 14, we replace the discontinuous functions $f_s(u_s)$ from (17.2) by smooth inside ε-vicinities of the discontinuity point functions $\overline{f}_s(u_s)$, as shown, for example, in Fig. 21 and 22. For the linear segment of the characteristics $\overline{f}_s(u_s)$ in Eqs. (17.1), it may be assumed that

$$f_{(p,1)} = -\frac{1}{\varepsilon} \kappa_{(p,p)} u_{(p,1)}, \tag{17.4}$$

where $\varepsilon \ll 1$, and $\kappa_{(p,p)}$ is a diagonal matrix.

Equations (17.1) and (17.3) can be written in block form:

$$\frac{d}{dt}v_{(r,1)} = -a_{(r,r)}\overline{f}_{(r,r)}(u_{(r,1)}) - a_{(r,p-r)}\overline{f}_{(p-r,p-r)}(u_{(p-r,1)}) + g_{(r,1)},$$

$$\frac{d}{dt}v_{(n-r,1)} = -a_{(n-r,r)}\overline{f}_{(r,r)}(u_{(r,1)})$$

$$-a_{(n-r,p-r)}\overline{f}_{(p-r,p-r)}(u_{(p-r,1)}) + g_{(n-r,1)}, \tag{17.5}$$

$$\frac{d}{dt}x_{(r,1)} = v_{(r,1)}, \qquad \frac{d}{dt}x_{(n-r,1)} = v_{(n-r,1)},$$

$$u_{(r,1)} = b_{(r,r)}v_{(r,1)} + b_{(r,n-r)}v_{(n-r,1)} + c_{(r,1)}, \tag{17.6}$$

$$u_{(p-r,1)} = b_{(p-r,r)}v_{(r,1)} + b_{(p-r,n-r)}v_{(n-r,1)} + c_{(p-r,1)}. \tag{17.7}$$

After excluding the variable $v_{(r,1)}$ from the system (17.5) and (17.7), it follows from (17.6) that:

$$v_{(r,1)} = b^{-1}_{(r,r)}(u_{(r,1)} - b_{(r,n-r)}v_{(n-r,1)} - c_{(r,1)}). \tag{17.8}$$

Substitution of (17.8) into (17.5) and (17.7) leads to:

$$b^{-1}_{(r,r)}\frac{d}{dt}u_{(r,1)} = -a_{(r,r)}\overline{f}_{(r,r)}(u_{(r,1)}) - a_{(r,p-r)}\overline{f}_{(p-r,p-r)}(u_{(p-r,1)})$$

$$+g_{(r,1)} - u_{(r,1)}\frac{d}{dt}b^{-1}_{(r,r)} + \frac{d}{dt}\left[b^{-1}_{(r,r)}(b_{(r,n-r)}v_{(n-r,1)} + c_{(r,1)})\right],$$

$$\frac{d}{dt}v_{(n-r,1)} = -a_{(n-r,r)}\overline{f}_{(r,r)}(u_{(r,1)})$$

$$-a_{(n-r,p-r)}\overline{f}_{(p-r,p-r)}(u_{(p-r,1)}) + g_{(n-r,1)}, \tag{17.9}$$

$$\frac{d}{dt}x_{(r,1)} = b^{-1}_{(r,r)}(u_{(r,1)} - b_{(r,n-r)}v_{(n-r,1)} - c_{(r,1)}),$$

$$\frac{d}{dt}x_{(n-r,1)} = v_{(n-r,1)},$$

$$u_{(p-r,1)} = b_{(p-r,r)}b^{-1}_{(r,r)}(u_{(r,1)} - b_{(r,n-r)}v_{(n-r,1)} - c_{(r,1)})$$

$$+b_{(p-r,n-r)}v_{(n-r,1)} + c_{(p-r,1)}.$$

It is assumed that the system (17.9) resides in a small ε-vicinity of the discontinuity points of functions (17.2) with respect to all u_s ($s = 1, \dots, r$). An additional normalization of these small variables in (17.9) should be done:

$$u_{(r,1)} = \varepsilon z_{(r,1)}, \tag{17.10}$$

$$\begin{aligned}
\varepsilon b^{-1}_{(r,r)} \frac{d}{dt} z_{(r,1)} &= -a_{(r,r)} \overline{f}_{(r,r)}(z_{(r,1)}) - a_{(r,p-r)} \overline{f}_{(p-r,p-r)}(u_{(p-r,1)}) \\
&\quad + g_{(r,1)} - \varepsilon z_{(r,1)} \frac{d}{dt} b^{-1}_{(r,r)} + \frac{d}{dt}\left[b^{-1}_{(r,r)}(b_{(r,n-r)} v_{(n-r,1)} + c_{(r,1)})\right], \\
\frac{d}{dt} v_{(n-r,1)} &= -a_{(n-r,r)} \overline{f}_{(r,r)}(z_{(r,1)}) \\
&\quad - a_{(n-r,p-r)} \overline{f}_{(p-r,p-r)}(u_{(p-r,1)}) + g_{(n-r,1)}, \\
\frac{d}{dt} x_{(r,1)} &= b^{-1}_{(r,r)}(\varepsilon z_{(r,1)} - b_{(r,n-r)} v_{(n-r,1)} - c_{(r,1)}), \\
\frac{d}{dt} x_{(n-r,1)} &= v_{(n-r,1)}, \\
u_{(p-r,1)} &= b_{(p-r,r)} b^{-1}_{(r,r)}(\varepsilon z_{(r,1)} - b_{(r,n-r)} v_{(n-r,1)} - c_{(r,1)}) \\
&\quad + b_{(p-r,n-r)} v_{(n-r,1)} + c_{(p-r,1)}.
\end{aligned} \tag{17.11}$$

The equation obtained from (17.6) by substitution of (17.10) should be added to (17.11):

$$\varepsilon z_{(r,1)} = b_{(r,n)} v_{(n,1)} + c_{(r,1)}. \tag{17.12}$$

The system (17.11) is singularly perturbed. After reduction to Cauchy form, we may use the Tikhonov theorem. The degenerate system for (17.11) is constructed by setting $\varepsilon = 0$:

$$\begin{aligned}
0 &= -a_{(r,r)} \overline{f}_{(r,r)}(\overline{z}_{(r,1)}) - a_{(r,p-r)} \overline{f}_{(p-r,p-r)}(\overline{u}_{(p-r,1)}) + g_{(r,1)} \\
&\quad + \frac{d}{dt}\left[b^{-1}_{(r,r)}(b_{(r,n-r)} \overline{v}_{(n-r,1)} + c_{(r,1)})\right], \\
\frac{d}{dt} \overline{v}_{(n-r,1)} &= -a_{(n-r,r)} \overline{f}_{(r,r)}(\overline{z}_{(r,1)}) \\
&\quad - a_{(n-r,p-r)} \overline{f}_{(p-r,p-r)}(u_{(p-r,1)}) + g_{(n-r,1)}, \\
\frac{d}{dt} \overline{x}_{(r,1)} &= b^{-1}_{(r,r)}(-b_{(r,n-r)} v_{(n-r,1)} - c_{(r,1)}), \\
\frac{d}{dt} \overline{x}_{(n-r,1)} &= \overline{v}_{(n-r,1)}, \\
\overline{u}_{(p-r,1)} &= b_{(p-r,r)} b^{-1}_{(r,r)}(-b_{(r,n-r)} \overline{v}_{(n-r,1)} - c_{(r,1)}) \\
&\quad + b_{(p-r,n-r)} \overline{v}_{(n-r,1)} + c_{(p-r,1)}.
\end{aligned} \tag{17.13}$$

Equation (17.12) with $\varepsilon = 0$ takes the form:

$$0 = b_{(r,n)} \overline{v}_{(n,1)} + c_{(r,1)}. \tag{17.14}$$

Equation (17.14) is, generally speaking, a nonintegrable expression for coordinates and velocities of the initial system. Therefore it determines a kinematic nonholonomic constraint realized by Coulomb friction forces. Reactions of these constraints, $f_{(r,1)}$, depending on inertial and other forces, are determined from the first equation. The magnitudes of $f_{(r,1)}$ correspond to the Coulomb hypothesis that all forces are balanced by friction forces when sliding is impossible. By excluding the reactions $f_{(r,1)}$ from the first two equations of (17.13), the equations of motion of a system with constraints are obtained.

Now the conditions of the Tikhonov theorem about the validity of limit passage from (17.11) to (17.13) should be verified.

Let us construct a system adjoined to (17.11). To that end, $dv_{(n-r,1)}/dt$ is excluded in the first two equations of (17.11), and in the resulting equation we pass $\tau = t/\varepsilon$ on to fast time. After calculations, the following equation is obtained

$$b^{-1}_{(r,r)} \frac{dz_{(r,1)}}{d\tau} = h_{(r,r)} f_{(r,1)} + \psi_{(r,1)}, \tag{17.15}$$

where the elements of the matrix $h_{(r,r)}$ depend on $x_1, \ldots, x_n = \text{const}$, and elements of the column $\psi_{(r,1)}$ depend on $x_1, \ldots, x_n; v_{r+1}, \ldots, v_n = \text{const}$. The equilibrium positions of the system (17.15) are determined by the system of equations

$$h_{(r,r)} f_{(r,1)} = -\psi_{(r,1)}. \tag{17.16}$$

System (17.16) is linear with respect to f_s. Its solution with respect to f_s is unique, if

$$|h_{(r,r)}| \neq 0. \tag{17.17}$$

Due to (17.2) and definition of the functions $\overline{f}_s$, the roots of the system (17.16) satisfy the restrictions

$$|f_s| < \kappa_s; \qquad s = 1, \ldots, r. \tag{17.18}$$

When conditions (17.17) and (17.18) are satisfied, the equilibrium states z_s^0 of the adjoined system (17.16) are obtained from the equation $f_s^0 = \overline{f}_s(z_s)$, using the calculated roots f_s^0 of the system (17.16). The mechanical sense of the restriction (17.18) is obvious; that is, Coulomb friction should not be over its limiting value.

To analyze the stability of the equilibrium position of z_s^0, the equations in deviations must be constructed using (17.15) and (17.4):

$$\Delta z_s = z_s - z_s^0.$$

Conditions of asymptotic stability of the Tikhonov theorem are satisfied if the system for z_s is located in the break zone $|z_s| < 1$ and if the roots of the characteristic determinant for (17.15), (17.4), and (17.10)

$$|b^{-1}_{(r,r)}\lambda + b_{(r,r)}\kappa_{(r,r)}| = 0,$$

lie in a complex plane to the left of an imaginary axis.

Equations in deviations Δz_s are the equations of mechanical systems under the action of only linear forces of friction. If dissipation in this system is complete, then asymptotic stability of its trivial solution is fulfilled according to conditions from Magnus [47].

The preceding conditions guarantee correctness of passage to a limit model of kinematic constraint only in a finite time interval. For correctness of passage in an infinite interval, according to Sec. 11.3, asymptotic stability of particular solutions of a limit system (17.13) with respect to initial perturbations is also required.

If this condition is not satisfied, the validity of the model of kinematic (nonholonomic) constraint on an infinite time interval becomes doubtful. Fairness of this doubt is convincingly confirmed to be true by the problem considered in Sec. 15.2.

Now it is assumed that some of the roots f_s, $s = 1, \ldots, q < r$, of the system (17.16) do not satisfy the condition (17.18). In this case the system (17.15) has no equilibrium positions with respect to corresponding variables z_s. This fact means that a mechanical system is not held in the stagnation zone with respect to these variables (that is, external forces, reduced to a given pair exceed the limit value of Coulomb friction in this pair). Then, for completing the definition of the system (17.5) for $s = 1, \ldots, q$, it should be assumed that $f_s = -\kappa_s \operatorname{sgn} u_s = \pm\kappa_s$, $s = 1, \ldots, q$, depending on the sign of u_s. Singular perturbations of $\varepsilon du_s/dt$ occur only in some equations for $s = q+1, \ldots, r$. Forming of appropriate equations and their analysis are carried out in the same way as in the previous case. When all the required conditions in the system are satisfied, kinematic (nonholonomic) constraints are realized only for several pairs with sliding for $s = q+1, \ldots, r$.

In the process of motion of the system (17.13) with constraints, the

coefficients and the right-hand sides of Eqs. (17.16) are changing in time. Therefore, their roots with respect to f_s are changing too. Different roots at different times may not satisfy the conditions (17.18). This leads to successive (as the system is moving) imposing and lifting of various constraints.

Thus, the most significant condition in the problem of kinematic constraints realization is the condition (17.18) meaning finiteness of the forces implementing a constraint. Without verifying this condition the model of kinematic constraints is as far incorrect as, for example, the model of unilateral holonomic constraints is incorrect without verifying retaining conditions on them.

Remark 1. Degenerate equations take especially simple form for $r = n$. Then $v_{(n-r,1)} = 0$ in (17.13) and all dynamic equations of the system transform to quasistatical equations of balance of forces.

Remark 2. At first sight it may seem that the equations of constraints (17.14) form a system inconsistent with a closed set of Eqs. (17.13). While discussing this doubt one should remember that equations (17.14) are approximate. Discrepancies of these equations are equal to εz_s by (17.12). When they are multiplied by the order coefficients $1/\varepsilon$ inside a break zone then generate terms $\overline{f}_s$ of the order of unity in (17.13). Following the principle of smallness hierarchy for the terms of different orders in (17.11), Eq. (17.12) should be considered as an irrelevant estimating relation.

Remark 3. Following a tradition of theoretical mechanics, the problem of limit passage to a model of nonholonomic constraints has been solved here under the assumption that holonomic constraints are already imposed and, therefore, passage to a limit similar to $K_s \to \infty$ from Sec. 16 has already been performed. We may expect that simultaneous limit passage to the forces providing both holonomic and nonholonomic constraints would lead to new situations. These situations are discussed in Sec. 22.2 using the example of a rolling deformable wheel.

17.2 Change of kinematic constraints for rolling of a braked wheel

As the simplest model of a car braking on an inclined plane, we consider a planar motion of a heavy uniform disk on an axis [39]. The equations of motion (see Fig. 34) have a form

$$M\frac{dV}{dT} = F(U) + G, \qquad I\frac{d\Omega}{dT} = RF(U) + L(\Omega),$$
$$U = V + R\Omega, \qquad F(U) = -F_0 \operatorname{sgn} U, \tag{17.19}$$
$$L(\Omega) = -L_0 \operatorname{sgn} \Omega.$$

Here M, R, I are the weight, the radius, and the moment of inertia of the

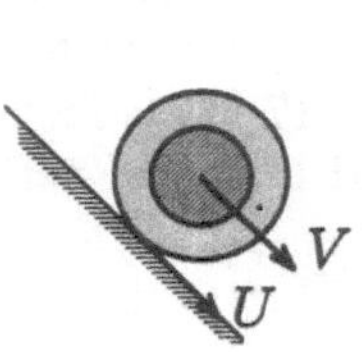

Fig. 34

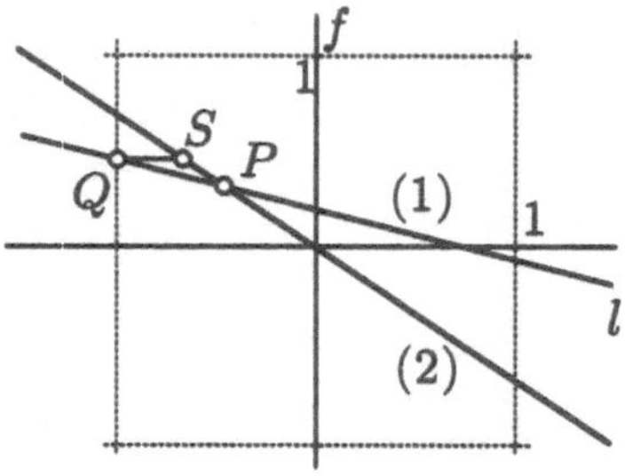

Fig. 35

disk, V is the velocity of the center of the mass, Ω is the absolute angular velocity, U is the velocity of a disk support point, $F(U)$ is the Coulomb friction in this point, and G is the component of gravity force along the supporting plane. In (17.19) a preliminary normalization

$$t = \frac{T}{T_*}, \qquad v = \frac{V}{V_*}, \qquad u = \frac{U}{U_*}, \qquad \omega = \frac{\Omega}{\Omega_*}.$$

should be done.

It is assumed that

$$\frac{MV_*}{T_*F_0} = 1, \quad , \frac{I\Omega_*}{T_*F_0R} = 1, \quad U_* = V_*, \quad I = Mr_0^2.$$

Then (17.19) takes the form:

$$\begin{aligned} \frac{dv}{dt} &= f(u) + g, \\ \frac{d\omega}{dt} &= f(u) + \lambda l(\omega), \\ u = v + \varrho\omega, \quad f(u) &= -\operatorname{sgn} u, \quad l(\omega) = -\operatorname{sgn}\omega. \end{aligned} \tag{17.20}$$

Here $g = G/F_0$, $\lambda = L_0/F_0R$, $\varrho = (R/r_0)^2$. For systems (17.20) the number of degrees of freedom n and the number of pairs with sliding p coincide: $p = n$. The passage in (17.20) to phase variables u, ω which are the arguments of discontinued functions leads to:

$$\begin{aligned} \frac{du}{dt} &= (1+\varrho)f(u) + \lambda\varrho l(\omega) + g, \\ \frac{d\omega}{dt} &= f(u) + \lambda l(\omega), \\ f(u) = -\operatorname{sgn} u, &\qquad l(\omega) = -\operatorname{sgn}\omega. \end{aligned} \tag{17.21}$$

According to the results of Sec. 17.1.1, a system similar to (17.13) can be constructed and analyzed:

$$(1+\varrho)f + \lambda\varrho l + g = 0, \qquad f + \lambda l = 0. \tag{17.22}$$

In Fig. 35 the straight lines (1) and (2) in a plane of variables f, l are determined by the first and second equations of (17.22), respectively. If the point $P(f_p^0 = -g, l_p^0 = g/l)$ of straight line intersection is located inside the strip $|f| < 1$, and at a certain time moment the variables of the system take the values $u = 0$, $\omega \neq 0$, then the further the system will move without sliding in the support point.

If the point P is located inside the strip $|l| < 1$, and the variables of the system take the values $u \neq 0$, $\omega = 0$, then further motion of the system consists of sliding without turning around the axis.

If the point P is located inside the square $|f|, |l| < 1$, and the variables take values $u = 0$, $\omega = 0$, then the disk will remain immovable.

As an example, let us consider motion such that at the initial time moment $u(0) = 0$, and $\omega(0) < 0$ is a finite magnitude. In this case, at least for a short interval of time the conditions $|u| < \varepsilon$, $\omega < 0$ are satisfied. A small parameter can occur in the equations of motion (17.21) only in the course of normalization of du/dt, and $l(\omega) = -\operatorname{sgn}\omega = +1$. The corresponding degenerate system has the form:

$$\begin{aligned} 0 &= (1+\varrho)f + \lambda\varrho\cdot 1 + g, \\ \frac{d\omega}{dt} &= f + \lambda\cdot 1. \end{aligned} \tag{17.23}$$

The quantities f, l in (17.23) are specified by a point Q of intersection of straight line (1) with the line $l = +1$. Development of the process with respect to the variable ω depends on the point P location. Let this point lie inside the square $|f|, |l| < 1$, as shown in Fig. 35. Then $d\omega/dt = \lambda(l_Q - l_S) > 0$ can be easily obtained from (17.23), where $l_Q - l_S$ is the difference of the abscissas of points Q, S (see Fig. 35). The magnitude of angular velocity will decrease, and in a certain time $\Delta t = |\omega(0)|/\lambda(l_Q - l_S)$ the system will stop with respect to both variables u and ω. It should be noted that at this moment friction forces jump from the values corresponding to the point Q to the values corresponding to the point P.

We assume that the point P is located not inside the square $|f|, |l| < 1$, but inside the horizontal strip $|f| < 1$. Thus, from (17.23) it can be easily obtained that $d\omega/dt < 0$, the absolute value of angular velocity increases, and the wheel cannot be braked at the inclined plane.

A more complicated example, when a three-wheeled carriage moves on a plane with the transition from one constraint to another, is discussed in Novozhilov [52].

17.3 Kinematic indefinability in a rolling rail car problem

The rail car is a complicated mechanical system possessing a great number of degrees of freedom. Its characteristic feature is a significant diversity of its natural frequencies. Dynamics of railway cars have been explored by methods of numerical analysis; the results are presented in Lazaryan et al. [53]. According to these data, variation of eigenfrequencies may achieve three decimal exponents. By this fact, numerical modeling of such a problem is an extremely hard task, and analytical exploration is not suitable because of the high order of the system. These arguments justify the urgency of simplified modeling of rail car dynamics.

17.3.1 We start with considering the simplest kind of car, that is, a car possessing a single wheel pair.

It is assumed that the car construction is absolutely rigid, except at small vicinities of the points where wheels contact rails. At these points "pseudo-sliding" is possible due to small deformations [48, 53]. A path is considered to be horizontal and rectilinear. An immovable frame of reference O_1xyz is fixed: the x-axis bisects the distance between rails and its direction coincides with the direction of car motion, the y-axis is directed to the left (Fig. 36). Small variation of the car body height is ignored, and motion of the body is considered to be planar. The position of the body is specified by its center of mass O coordinates X, Y and turn angle Ψ.

The equations of motion for a "body+wheel pair" system are composed, assuming that longitudinal velocity is constant: $V_x = V$, and ignoring nonlinear terms with respect to Y, Ψ [48, 53],

$$\begin{aligned} M\frac{dV_y}{dT} &= F_{1y} + F_{2y} + F_y, & \frac{dY}{dT} &= V_y, \\ I\frac{d\Omega}{dT} &= F_{2x}l - F_{1x}l + M_z, & \frac{d\Psi}{dT} &= \Omega. \end{aligned} \tag{17.24}$$

Here, in addition to the previously defined notation I is the system moment of inertia relative to a vertical axis passing through the center of the mass, and $2l$ is the width of the track. Finally, (F_{1x}, F_{1y}), (F_{2x}, F_{2y}) stand for the projections onto the x, y axis of so-called creep forces, that is, the forces of interactions of wheels with rails, and F_y, M_z are the principal vector and principal moment of perturbing forces. (The wheels are enumerated as shown in Fig. 36.) A typical experimental characteristic of the creep forces is represented [48, 53] in Fig. 37. The zone of its linearity $u_x \ll 1$ is small, and the steepness coefficient is great: $K \approx F_0/\varepsilon$. Having passed the zone of linearity, this characteristic approaches saturation, quite similar to characteristics of forces $\overline{f}_s(u_s)$ from (17.5). Therefore, the system (17.24) is a special case of systems similar to (17.5).

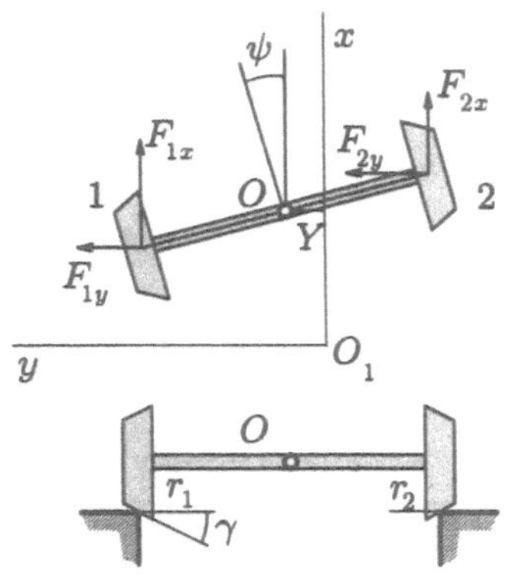

Fig. 36

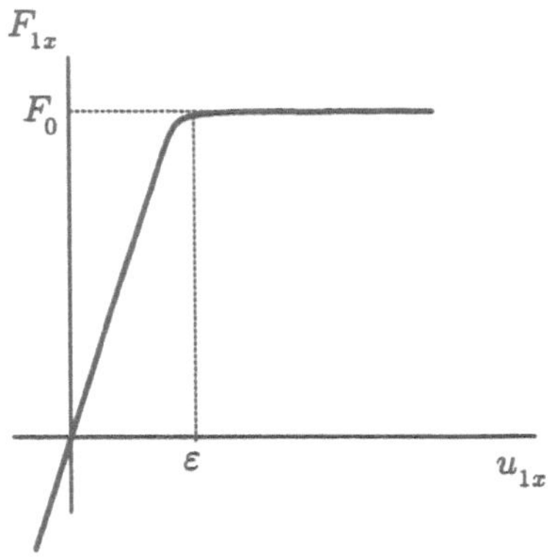

Fig. 37

Expressions defining creep forces within the zone of their characteristic linearity are, according to Neymark and Fufaev [48] and Lazaryan et al. [53]:

$$F_{1x} = -Ku_{1x}, \; F_{1y} = -Ku_{1y}, \; F_{2x} = -Ku_{2x}, \; F_{2y} = -Ku_{2y}. \tag{17.25}$$

Here K is the creep coefficient and $u_{1x}, \ldots, u_{2y}$ is the "pseudo-sliding" at the contact point. Magnitudes $u_{1x}, u_{2x}, u_{1y}, u_{2y}$ are dimensionless projections V_{1x}, V_{1y} and V_{2x}, V_{2y} corresponding to each of the wheels and specifying velocities of contact points relative to the rails:

$$u_{1x} = V_{1x}/V, \quad \ldots, \quad u_{2y} = V_{2y}/V. \tag{17.26}$$

We calculate by the velocity summation theorem:

$$V_{1x} = V_{1x}^e + V_{1x}^r, \quad \ldots, \quad V_{2y} = V_{2y}^e + V_{2y}^r. \tag{17.27}$$

Reference-frame velocities are:

$$V_{1x}^e = V - l\Omega, \qquad V_{2x}^e = V + l\Omega, \qquad V_{1y}^e = V_{2y}^e = V_y. \tag{17.28}$$

Relative velocities are:

$$\begin{aligned} V_{1x}^r &= -r_1\omega_n, & V_{2x}^r &= -r_2\omega_n, & V_{1y}^r &= -r_1\omega_n\Psi, \\ V_{2y}^r &= -r_2\omega_n\Psi, & r_1 &= r + Y\gamma, & r_2 &= r - Y\gamma. \end{aligned} \tag{17.29}$$

Here r_1, r_2 is the distance between the axis of the corresponding wheel and its contact point, r is the mean radius of the wheel, γ is its conicity, and ω_n is the angular velocity of the wheel pair. It should be noted that $r\omega_n = V$.

Having accomplished transformations of (17.26) - (17.29), we obtain an approximation:

$$\begin{aligned} u_{1x} &= (-l\Omega - \frac{V\gamma}{r}Y)\frac{1}{V}, & u_{1y} &= (V_y - V\Psi)\frac{1}{V}, \\ u_{2x} &= (l\Omega + \frac{V\gamma}{r}Y)\frac{1}{V}, & u_{2y} &= (V_y - V\Psi)\frac{1}{V}. \end{aligned} \tag{17.30}$$

The system of equations (17.24), (17.25), (17.30) is closed. Having imposed kinematical constraints, that is, conditions for nonsliding at the contact points, we obtain according to (17.30) four equations:

$$u_{1x} = 0, \quad u_{1y} = 0, \quad u_{2x} = 0, \quad u_{2y} = 0. \tag{17.31}$$

They are relations between two coordinates Y and Ψ. However, it is obvious from (17.30) that within adopted linear approximation $u_{1x} = -u_{2x}$ and $u_{1y} = u_{2y}$. Hence, formally imposed on the system constraints Eqs. (17.31) generate only two equations. They can be written, after taking into account kinematical equations from (17.24), in the form:

$$\begin{aligned} &\frac{dY}{dT} - V\Psi = 0, \\ &l\frac{d\Psi}{dT} + \frac{V\gamma}{r}Y = 0. \end{aligned} \tag{17.32}$$

Equations (17.32) define oscillatory motion with respect to each variable. Exclusion of, for example, Ψ, leads to:

$$\frac{d^2Y}{dT^2} + \omega_0^2 Y = 0, \qquad \omega_0^2 = \frac{V^2\gamma}{rl}. \tag{17.33}$$

These oscillations of a rail car are sometimes called "kinematical wigging."

Two features of formal model (17.32) based on (17.31) should be noted. First, nothing is known of its correctness, and, secondly, it does not permit us to to estimate the influence of perturbations. Now a simplified limit model is constructed using the fractional analysis technique.

17.3.2 Let us introduce in (17.30) the notation:

$$\begin{aligned} &U_y = V_y - V\Psi, \\ &U_\psi = l\Omega + \frac{V\gamma}{r}Y. \end{aligned} \tag{17.34}$$

Using (17.34), in (17.24), (17.25), and (17.30) the variables V_y, Ω are excluded. The equations of motion in variables U_y, U_ψ, Y, Ψ may be written as:

$$\begin{aligned} &M\frac{dU_y}{dT} = -\frac{2K}{V}U_y + F_y - \frac{MV}{l}\left(U_\psi - \frac{V\gamma}{r}Y\right), \\ &\frac{I}{l}\frac{dU_\psi}{dT} = -\frac{2K}{V}U_\psi l + M_z + \frac{IV\gamma}{rl}(U_y + V\Psi), \\ &\frac{dY}{dT} = U_y + V\Psi, \qquad \frac{d\Psi}{dT} = \frac{1}{l}\left(U_\psi - \frac{V\gamma}{r}Y\right). \end{aligned} \tag{17.35}$$

Equations (17.35) should be normalized:

$$u_y = \frac{U_y}{U_*}, \quad u_\psi = \frac{U_\psi}{U_*}, \quad y = \frac{Y}{Y_*}, \quad \psi = \frac{\Psi}{\Psi_*},$$
$$f_y = \frac{F_y}{F_*}, \quad m_z = \frac{M_z}{M_*}, \quad t = \frac{T}{T_*}. \tag{17.36}$$

The performance of this operation here is not identical to that demonstrated in Sec. 17.1. In Sec. 17.1 we have dealt with the discontinuous characteristic of Coulomb friction. The small parameter $\varepsilon \to 0$ was introduced by formal completion of a definition of the discontinuous function inside its break zone. In the problem now being considered, coefficient K of the creep forces takes finite value. This should be taken into account while performing normalization. According to the data of Lazaryan et al. [53], we set $U_* = Y_*/T_* = V\Psi_*$, and assume that $2KU_*/V = F_*$, $M_* = F_*l$. Let the characterizing time of slow motions being selected be $T_* = T_0$; $T_0 = 1/\omega_0$ is a time constant for kinematical wigging (17.33).

After normalization the system (17.35) takes the form:

$$\mu \frac{du_y}{dt} = -u_y + f_y - \mu(\frac{1}{\tau_l}u_\psi - y), \qquad \frac{dy}{dt} = u_y + \psi,$$
$$\mu\lambda\frac{du_\psi}{dt} = -u_\psi l + m_z + \mu\lambda\tau_l(u_y + \psi), \quad \frac{d\psi}{dt} = \frac{1}{\tau_l}(u_\psi - y). \tag{17.37}$$

Here $\mu = T_1/T_0$, $T_1 = MV/2K$, $\tau_l = \sqrt{\gamma l/r}$, $\lambda = (\rho_0/l)^2$, and ρ_0 is the radius of inertia. After substitution of numerical data from Lazaryan et al. [53] we obtain $\mu \ll 1$, $\tau_l \sim 1$, $\lambda \sim 1$.

System (17.37) is singularly perturbed. The corresponding degenerate system of equations is:

$$0 = -\overline{u}_y + f_y, \quad \frac{d\overline{y}}{dt} = \overline{u}_y + \overline{\psi},$$
$$0 = -\overline{u}_\psi + m_z, \quad \frac{d\overline{\psi}}{dt} = \frac{1}{\tau_l}\overline{u}_\psi - \overline{y}. \tag{17.38}$$

The orrectness of using Eqs. (17.38) is verified in the same way as in the final part of Sec. 17.1. The only significant is condition is that they must be located inside the zone of the creep characteristic linearity. Therefore, Eqs. (17.38) may be assumed to define the limit model.

It should be noted that this model is valid only in the finite time interval because for (17.38) Gradstein conditions do not hold.

Equations (17.38) are rather similar to Eqs. (17.32), obtained by imposing kinematic nonholonomic constraints (17.31). Therefore, differential equations from (17.38) can be treated as the equations of nonholonomic constraints. From finite equations (17.38), the principal vector ($\overline{u}_y$) and the principal moment ($\overline{u}_\psi$) of reactions of these constraints can be determined.

By using (17.36), let us pass in (17.38) to the initial dimensional variables:

$$0 = -\frac{2K}{V}U_y + F_y, \qquad \frac{dY}{dT} = U_y + V\Psi,$$
$$0 = -\frac{2K}{V}U_\psi l + M_z, \qquad \frac{d\Psi}{dT} = -\frac{1}{l}(U_\psi - \frac{\gamma V}{r}Y). \tag{17.39}$$

The error of approximating solutions of (17.35) by the system (17.39) with respect to variables Y, Ψ is also of the order of μ. The frequency of natural oscillations for (17.39) coincides, of course, with the value of ω_0 from (17.33).

17.3.3 Let us consider a more complicated model of a rail car: a two-axis wagon (see Fig. 38) [54].

The notation of Fig. 36 remains valid here and, in addition, $2a$ designates the distance between axes; wheels are enumerated by 1,2,3,4. Let

$$F_{1x} = -Ku_{1x}, \quad \ldots, \quad F_{4y} = -Ku_{4y} \tag{17.40}$$

be the corresponding projections of creep forces in the points of contact of the wheels with the rails. By analogy with (17.30), the linear expressions of components (17.40) corresponding to relative sliding at the contact points

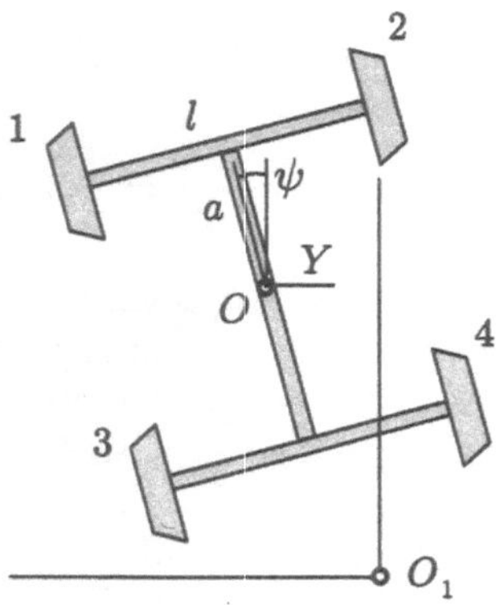

Fig. 38

are:

$$
\begin{aligned}
u_{1x} &= \Big[-l\Omega - \frac{V\gamma(Y+a\Psi)}{r}\Big]\frac{1}{V}, & u_{1y} &= (V_y + a\Omega - V\Psi)\frac{1}{V},\\
u_{2x} &= \Big[l\Omega + \frac{V\gamma(Y+a\Psi)}{r}\Big]\frac{1}{V}, & u_{2y} &= (V_y + a\Omega - V\Psi)\frac{1}{V},\\
u_{3x} &= \Big[-l\Omega - \frac{V\gamma(Y-a\Psi)}{r}\Big]\frac{1}{V}, & u_{3y} &= (V_y - a\Omega - V\Psi)\frac{1}{V},\\
u_{4x} &= \Big[l\Omega - \frac{V\gamma(Y-a\Psi)}{r}\Big]\frac{1}{V}, & u_{4y} &= (V_y - a\Omega - V\Psi)\frac{1}{V}.
\end{aligned}
\tag{17.41}
$$

An attempt to impose constraints on the system, that is, to prohibit sliding leads to a system of equations $u_{1x} = \ldots = u_{4y} = 0$, possessing only a trivial solution $\Omega, V_y, \Psi, Y \equiv 0$.

Let us compose equations of the limit model using the results of Sec. 17.1. In the problem under consideration $n = 2$ and $p = 8$; hence, the system is kinematically indefinable. For (17.41), let us write out a dimensional matrix $B_{(8,2)}$ similar to the matrix $b_{(p,n)}$ of (17.3). The rank of $B_{(8,2)}$ equals two. Thus, $r = n = 2$ and, by analogy with (17.39), the limit model is formed by quasistatical and kinematical equations.

The following calculations in many respects repeat Sec. 17.3.2.

The equations of motion of a two-axis car differ from (17.24) only by the principal vector and the principal moment of contact forces.

$$
\begin{aligned}
M\frac{dV_y}{dT} &= F_{1y} + \ldots + F_{4y} + F_y = 0, \qquad \frac{dY}{dT} = V_y,\\
I\frac{d\Omega}{dT} &= l(F_{2x} + F_{4x}) - l(F_{1x} + F_{3x})\\
&\quad + a(F_{1y} + F_{2y}) - a(F_{3y} + F_{24}) + M_z,\\
\frac{d\Psi}{dT} &= \Omega.
\end{aligned}
\tag{17.42}
$$

It is supposed that the arguments $u_{1x}, \ldots, u_{4y}$ of all creep force components in (17.42) are inside zones of their characteristics linearity. Substituting (17.41) and (17.40) into (17.42) produces

$$
\begin{aligned}
M\frac{dV_y}{dT} &= -\frac{4K}{V}(V_y - V\Psi) + F_y, & \frac{dY}{dT} &= V_y,\\
I\frac{d\Omega}{dT} &= -\frac{4K}{V}\Big[(l^2 + a^2)\Omega + \frac{V\gamma l}{r}Y\Big] + M_z, & \frac{d\Psi}{dT} &= \Omega.
\end{aligned}
\tag{17.43}
$$

Let the fast variables of (17.43) be

$$
U_y = (V_y - V\Psi), \qquad U_\psi = \Big[(l^2 + a^2)\Omega + \frac{V\gamma l}{r}Y\Big]. \tag{17.44}
$$

Now using (17.44), the variables V_y, Ω should be substituted in (17.43) by U_y, U_ψ, and the obtained equations should be normalized according to (17.36) and a system of the same structure as (17.37) which is singularly perturbed with respect to the parameter $\mu = T_1/T_*$. Unlike (17.37), here $T_1 = MV/4K$, and T_* is the time constant of slow motion components, which is not yet specified.

Let us write down the degenerate system of equations with respect to μ similar to (17.38) and let us pass to the dimensional variables as in (17.39):

$$\begin{aligned} 0 &= -\frac{4K}{V}U_y + F_y, & \frac{dY}{dT} &= U_y + V\Psi, \\ 0 &= -\frac{4K}{V}U_\psi + M_z, & \frac{d\Psi}{dT} &= \frac{1}{(l^2+a^2)}(U_\psi - \frac{V\gamma l}{r}Y). \end{aligned} \tag{17.45}$$

By analogy with Sec. 17.3.2, the differential equations in (17.45) can be considered to be equations of nonholonomic constraints; $-4KU_y/V$ and $-4KU_\psi/V$ are the principal vector and the principal moment of reactions of these constraints.

The frequency of the natural oscillations of the system (17.45) is determined by equality

$$\omega_0^2 = \frac{V^2\gamma l}{(l^2+a^2)r}.$$

Magnitude ω_0 is the frequency of the kinematical wiggings of a two-axis car. While normalizing the equations of the car, it is necessary to choose $T_* = T_0 = 1/\omega_0$. It uniquely specifies the magnitude $\mu = T_1/T_0$ of the small parameter problem.

In Novozhilov [54] a more detailed analysis of a two-axis car has been accomplished. Dynamics of wheel pairs and dynamics of a longitudinal motion has been accomplished. A hierarchy of motion components developing in four time scales has been established. Separating motion components has been carried out by means of results referred to in Remark 2 of Sec. 11.2. An analogous method may be used in separating the of high-frequency elastic components of motion specified by the final elasticity of the car. If these frequencies are much greater than the frequency of kinematical wiggings, then the final elasticity of the car does not influence the limit model features.

By analogy with the system (17.38), the natural motions of the systems (17.45) are undamped. As Gradstein conditions do not hold, the model (17.45) is valid only in the finite time interval.

The limit model of a car motion becomes much more complicated if restrictions on lateral motions of wheel pairs caused by the flanges are taken into account. Force interactions with the rail structure are elastic ones characterized by high stiffness [53]. An approximate model of the car can be constructed by simultaneous limit passage by stiffness of elastic forces as in Sec. 16, and by factors of creep forces, as in Sec. 17.

This limit model possesses a variable structure and is specified by a system of equations of a different order depending on number and list of the wheels contacting a flange.

18 Limit model of servoconstraint

18.1 Conditions of servoconstraint realizability

The notion "servoconstraint" arises only in the theory of controlled mechanical systems. The servoconstraints are the constraints, imposed on the phase variables of the system and realized with high accuracy due to "hard" control. The control stiffness is provided by the "large" values of the control coefficients or due to relay control when in the domain of discontinuity the system can be considered linear with the infinite coefficient of proportionality. An example of the servoconstraint arising is the problem concerned with angular motion control of a spacecraft from Sec. 15.1. For this problem, the representative point of the system is retained on the interval $[S_1, S_2]$ (Fig. 24) by the relay control. So the equation $u = \varphi + \tau_0\omega = 0$ can be considered as the servoconstraint equation.

The main difference of the servoconstraints from ordinary holonomic and nonholonomic constraints is the fact that the latter are realized by means of direct contact interactions of material solids. So in mechanics the reaction forces realizing the constraint always conform with the Third Newton's Law. While the system moves in accordance with the servoconstraint, one cannot indicate the force by means of which it acts upon the system. It is obvious, for example, for a spacecraft problem dealing with only one solid.

While analyzing a system with a servoconstraint, two problems arise:

- what servoconstraints should be chosen, and
- what should be the control realizing these servoconstraints.

The choice of servoconstraints is defined by the purpose of the concrete considered system — by its goals. The servoconstraint at this stage usually occurs as a "semiinverse setting of a problem," when some variables are assumed to be predetermined functions of time. Let us suppose that the choice of needed servoconstraints is performed beforehand and now deal with the problem of their realization.

The servoconstraints problem, obviously, is similar to the settings of previous sections; therefore, this problem solved analogously.

The normalized equations of the controlled mechanical system can be written in the form (17.1):

$$\frac{d}{dt}v_{(n,1)} = a_{(n,r)}f_{(r,1)} + g_{(n,1)}, \qquad \frac{d}{dt}x_{(n,1)} = v_{(n,1)}. \tag{18.1}$$

where f_s $(s = 1, \ldots, r)$ are the control forces implementing the servoconstraints.

It is assumed that the servoconstraints for the system (18.1) were chosen previously. Their equations are:

$$\begin{aligned} u_1 &= b_{11}v_1 + \ldots + b_{1n}v_n + c_1 = 0, \\ &\ldots \\ u_r &= b_{r1}v_1 + \ldots + b_{rn}v_n + c_r = 0, \end{aligned} \tag{18.2}$$

where $b_{11}, \ldots, b_{rn}; c_1, \ldots, c_r$ are defined and depend on $x_1, \ldots, x_n, t$.

The following limitation of generality in (18.2) should be noted: the servoconstraint depends linearly on the velocities v_k, the number r of constraints in (18.2) is assumed to coincide with the number of the controls in (18.1), and the number of servoconstraints does not exceed the number of the degrees of freedom $(r \leq n)$.

The coefficients in (18.2) are required to be smooth as needed in the future, and the rank of the matrix $[b_{sk}] \equiv b_{(rn)}$ is required to be equal to r for the whole domain of definition of the system (18.1). For definiteness, by analogy with Sec. 17.1, it is supposed that

$$|b_{(r,r)}| \neq 0. \tag{18.3}$$

It is required that Eqs. (18.2) be satisfied due to control with predefined accuracy. The quantities u_s are assumed to be small. Let us convolve relations (18.2) to the matrix equation, which is analogous to (17.3):

$$u_{(r,1)} = b_{(r,n)}v_{(n,1)} + c_{(r,1)}. \tag{18.4}$$

Let us suppose that the control system is equipped by the measuring devices for all phase variables $x_1, \ldots, x_n; v_1, \ldots, v_n$. The values $b_{(r,n)}$, $c_{(r,1)}$

Holonomic and Servo Constraints

are defined in (18.2), so the values $u_{(r,1)}$ can be calculated in accordance with (18.4) for any time moment. Here the errors of the measurements are not taken into account. They can be easily introduced into (18.4), if it is necessary.

Let us consider a control with negative feedback. It is assumed that the control is inertialess. Only a linear control scheme with respect to signal deviation is considered:

$$f_{(r,1)} = -K\Pi_{(r,r)}u_{(r,1)}. \tag{18.5}$$

Here K is the "large" coefficient, which provides the stiffness of the control and $\Pi_{(r,r)}$ is the square normalized ($\|\Pi\| \sim 1$) matrix, depending on $x_1, \ldots, x_n, t$. By means of this matrix the resetting of signal deviations $u_1, \ldots, u_r$ for controls $f_1, \ldots, f_r$ is described. For the applied problems, the device, which realizes the matrix, is called the coordinate transformer [33, 55].

The goal is stated of choosing the value K and the matrix $\Pi_{(r,r)}$, which provide the realization of the servoconstraints (18.2) with prescribed accuracy.

Let us estimate the required K. By virtue of preliminary normalization of the system (18.1), the following relations are valid: $\|a_{(n,r)}\| \sim 1$, $\|g_{(n,1)}\| \sim 1$.

It is assumed that the forces $g_{(n,1)}$ are slow. Then the norm of matrix $f_{(r,1)}$ is characterized by the static estimate $\|f_{(r,1)}\| \sim 1$. The numerical measure of the acceptable deviations of the system from servoconstraints is given as $\varepsilon \ll 1$; therefore, $\|u_{(r,1)}\| \sim \varepsilon$. Then from (18.5) with $\|\Pi_{(r,r)}\| \sim 1$ it follows that $K \sim 1/\varepsilon$. Further it is assumed that $K = 1/\varepsilon$.

Now the system (18.1), (18.4), and (18.5) coincides by form with the system (17.1), (17.3), and (17.4). The difference is in the fact that matrix $\kappa_{(p,p)}$ in (17.4) is diagonal with known elements, defined by the initial mechanical setting in (17.2). The corresponding matrix $\Pi_{(r,r)}$ in (18.5) is an arbitrary square matrix to be defined.

The following actions repeat the operations of Sec. 17.1. For the system (18.1), (18.4), and (18.5) the system is constructed in the form of (17.5). The change $z_s = u_s/\varepsilon$ transforms the system to the singularly perturbed form (17.11). For (17.11) the degenerated system is constructed as (17.13).

Now Eq. (17.14) should be treated as the equation of servoconstraints imposed on the system. The values $f_{(r,1)}$ from (17.13) define the values of the control forces, corresponding to the motion along servoconstraints. If in (17.13) the controls $f_{(r,1)}$ are excluded, then the obtained equations describe the motion of the system along servoconstraints under the action of other forces.

Let us consider the conditions of the validity of the limit transition to the equations of the controlled system similar to (17.13). The condition

$|h_{(r,r)}| \neq 0$ from (17.17), concerning the existence of the equilibrium positions f_s^0 of the system (17.15) with respect to $f_{(r,1)}$, does not depend on matrix $\Pi_{(r,r)}$. It is possible to influence this condition only by choosing the set of the controls $f_1, \ldots, f_r$ used for realization of the servoconstraints, if, of course, the system has such redundancy of the controls. This analysis can be performed for the defined concrete system. In the following it is assumed that the condition (17.17) is satisfied, and the roots of the system (17.16) with respect to f_s exist.

Conditions such as (17.18) can be applied for checking the entry of the control to restrictions, because the real control actions are always limited by an absolute value. The condition of equilibrium positions with respect to z_s^0 existence for the adjoined system follows from the equation similar to (18.5), with $f_s = f_s^0$:

$$|\Pi_{(r,r)}| \neq 0. \tag{18.6}$$

This condition, as opposed to Sec. 17.1, is nontrivial.

Let us estimate the equilibrium position stability of the adjoined system as (17.15), (17.4), and (17.10). This is a linear system with constant coefficients, because the slow variables x_k, t are considered to be constants. Let us write the characteristic matrix

$$M(\lambda) = b_{(r,r)}^{-1}\lambda + h_{(r,r)}\Pi_{(r,r)}. \tag{18.7}$$

The requirement of asymptotic stability of the Tikhonov theorem is satisfied if the roots of the characteristic equation $|M(\lambda)| = 0$ lie to the left of the imaginary axis. This requirement provides the stability of the system with respect to fast motions during stiff servoconstraint tracking. It is constructive, because it allows us to define the matrix $\Pi_{(r,r)}$ of the coordinate transformation according to (18.7). This problem is solved nonuniquely. However, the form of the matrix $\Pi_{(r,r)}$, which certainly solves the problem, may be defined quite easily. Let us define it by the relation

$$\Pi_{(r,r)} = h_{(r,r)}^{-1} b_{(r,r)}^{-1}. \tag{18.8}$$

Here $|h_{(r,r)}| \neq 0$ according to (17.17). Then (18.7) is transformed into

$$M(\lambda) = b_{(r,r)}^{-1}[\lambda E + E]. \tag{18.9}$$

Let us write the characteristic equation using (18.9)

$$|M(\lambda)| = |b_{(r,r)}^{-1}||\lambda E + E| = 0. \tag{18.10}$$

According to (18.3), the matrix $b_{(r,r)}$ is nonsingular; so from (18.10) it follows that

$$|\lambda E + E| = 0.$$

The roots of this equation are $\lambda_s = -1$ for all $s = 1, \dots, r$; this provides satisfaction of the Tikhonov theorem conditions concerning asymptotic stability.

The matrix $\Pi_{(r,r)}$ from (18.8) also satisfies the condition (18.6).

Thus, all conditions for validity of the transition to the equation similar to (17.13) are satisfied, and they can be considered as the equations of the system with the servoconstraint limit model.

18.2 Realization of servoconstraints, defining the manipulator extremity motion

Figure 39 shows the planar model of the robot extremity. If the grip is located in point C and the control moments are defined in the joints O, O_1, then Fig. 39 shows the robot "arm." If the joint in point C links an extremity and body of a vehicle, and the control moments are defined in the joints C and O_1, then Fig. 39 represents the "leg" of a legged vehicle during the support stage. For the definiteness let us consider the latter case [56]. It is assumed for simplicity that the elements 1 and 2 indicated in the figure are absolutely rigid and weightless, and the mass center of the vehicle is located at point C. The angular motions of the body are not considered.

Let us form the equations of the system motion using the equation of the change of momentum for the whole system, kinematic equations, and static equations of the moments for the inertialess systems "element 1 + element 2" and "element 1" with respect to the points C and O_1 respectively, and geometric relations

$$
\begin{gathered}
M\frac{d}{dT}\begin{bmatrix} V_x \\ V_y \end{bmatrix} = \begin{bmatrix} R_x \\ R_y \end{bmatrix} + \begin{bmatrix} 0 \\ -Mg \end{bmatrix}, \\
\frac{d}{dT}\begin{bmatrix} X \\ Y \end{bmatrix} = \begin{bmatrix} V_x \\ V_y \end{bmatrix}, \\
Q_1 + R_xY_1 - R_yX_1 = 0, \qquad Q + R_xY - R_yX = 0, \\
X_1 = -L_1 \sin\Phi_1, \qquad Y_1 = L_1\cos\Phi_1, \\
X = -L_1\sin\Phi_1 - L_2\sin\Phi_2, \qquad Y = L_1\cos\Phi_1 + L_2\cos\Phi_2.
\end{gathered} \tag{18.11}
$$

Here X, Y; X_1, Y_1 are coordinates of the points C and O_1 in a coordinate frame, indicated in Fig. 39. It also contains some indicated designations from (18.11). R_x, R_y are projections of the vector of support reaction at the point O, and Q_1, Q are the control moments at the joints O_1 and C.

Let us define as the desired program mode the motion with constant speed $V_x = V$ and the constant height $Y = H$. The smoothness of the

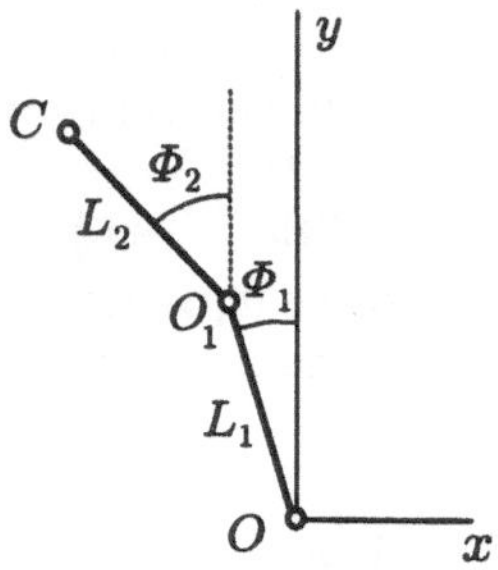

Fig. 39

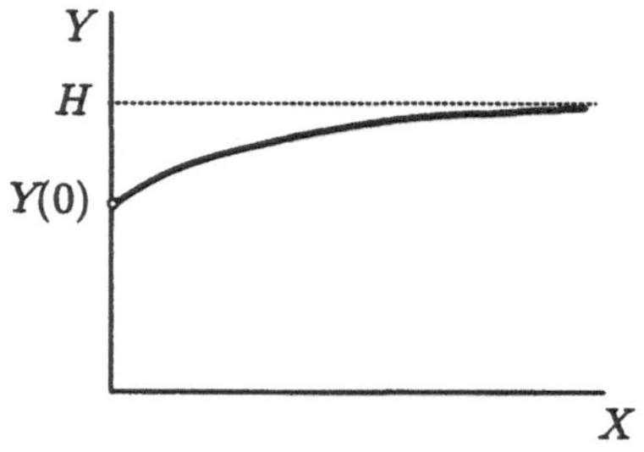

Fig. 40

passing to the program motion from an arbitrary initial point should be provided by the servoconstraints

$$U_x = 0, \qquad U_y = 0, \tag{18.12}$$

where

$$U_x = V_x - V, \qquad U_y = (V_y - 0) + \frac{1}{T_0}(Y - H). \tag{18.13}$$

Here T_0 = const is the dimensional parameter of time, adjusting the dimensions of the additives in (18.13). Equations (18.12) and (18.13), complemented by the kinematics equations from (18.11), are integrated and provide the motion along the servoconstraint

$$V_x^0 = V, \qquad Y^0 = H + (Y(0) - H)e^{-T/T_0}. \tag{18.14}$$

The trajectory of motion (18.14) is represented in Fig. 40. It can be seen that the servoconstraints (18.12) provide the smooth passing of the system to the program motion.

Let us choose the control which realizes the motion along the servoconstraints. Let us define the control in class (18.5). Then

$$\begin{bmatrix} Q_1 \\ Q \end{bmatrix} = -K\Pi \begin{bmatrix} U_x \\ U_y \end{bmatrix}. \tag{18.15}$$

Here K is the "large" coefficient, Π is the square matrix of the coordinate transformation, and U_x, U_y are deviations from servoconstraints defined by (18.13).

Let us exclude from (18.11) and (18.15) variables R_x, R_y; Q_1, Q. Let us perform the transition by means of (18.13) from the set of phase variables X, Y, V_x, V_y to the set X, Y, U_x, U_y, which contains deviations from servoconstraints. Thus, we obtain the system

$$
\begin{aligned}
M\frac{d}{dT}\begin{bmatrix} U_x \\ U_y \end{bmatrix} &= -KA\Pi\begin{bmatrix} U_x \\ U_y \end{bmatrix} + \begin{bmatrix} 0 \\ -Mg - \frac{M}{T_0^2}(Y-H)+\frac{M}{T_0}U_y \end{bmatrix}, \\
\frac{d}{dT}\begin{bmatrix} X \\ Y \end{bmatrix} &= \begin{bmatrix} U_x \\ U_y \end{bmatrix} + \begin{bmatrix} 0 \\ -\frac{1}{T_0}(Y-H) \end{bmatrix}, \\
A &= \frac{1}{\Delta}\begin{bmatrix} X - X_1 \\ y - Y_1 \end{bmatrix}, \qquad \Delta = YX_1 - XY_1.
\end{aligned}
\tag{18.16}
$$

Let us normalize the system (18.16), complemented by the geometric equations from (18.11). Let us join, as earlier while solving the concrete problems in Sec. 15.2 and 17.3.2, the stages of the preliminary normalization and normalization introducing the small parameter

$$
t = \frac{T}{T_*}, \quad x = \frac{X}{X_*}, \quad x_1 = \frac{X_1}{X_{1*}}, \quad y = \frac{Y}{Y_*}, \quad y_1 = \frac{Y_1}{Y_{1*}},
$$

$$
l_1 = \frac{L_1}{L_{1*}}, \quad l_2 = \frac{L_2}{L_{2*}}, \quad u_x = \frac{U_x}{U_*}, \quad u_y = \frac{U_y}{U_*}, \quad \varphi_1 = \frac{\Phi_1}{\Phi_*}, \quad \varphi_2 = \frac{\Phi_2}{\Phi_*}.
$$

The stride is assumed to be long. Let us accept $X_*=Y_*=L_*=X_{1*}=Y_{1*}=H$. Then for the matrix A from (18.16) it follows that $\|A\| \sim 1/H$.

In the first equation (18.16) the matrix Π should be defined so that matrix $A\Pi$ would be dimensionless. Let us require the satisfaction of $\|A\Pi\| \sim 1$, and choose the characterizing time to be the time of one step order: $T_* = X_*/V = H/V$. Let us assume the characterizing values of the control actions to be the values of the system weight order:

$$
KU_* = Mg. \tag{18.17}
$$

After the preceding normalization the system (18.16) takes the form:

$$
\begin{aligned}
\mu\frac{d}{dt}\begin{bmatrix} u_x \\ u_y \end{bmatrix} &= -A\Pi\begin{bmatrix} u_x \\ u_y \end{bmatrix} + \begin{bmatrix} 0 \\ -1-h(y-1) + \mu u_y/\tau_0 \end{bmatrix}, \\
\frac{d}{dT}\begin{bmatrix} x \\ y \end{bmatrix} &= \mu\gamma\begin{bmatrix} u_x \\ u_y \end{bmatrix} + \begin{bmatrix} 1 \\ -\frac{1}{\tau_0}(y-1) \end{bmatrix}.
\end{aligned}
\tag{18.18}
$$

In (18.18) dimensionless designations are introduced: $\mu=T_1/T_*$, $T_1=M/K$, $h = H/T_0^2 g$, $\tau_0 = T_0/T_*$, and $\gamma = T_* g/V$. Here T_1 is the small time constant of the system defined by the high stiffness K of the control.

The class of motion, which satisfies the following conditions: $\mu \ll 1$, h, τ_0, and $\gamma \sim 1$, is being considered. The characteristic error of the deviation of the system from the servoconstraints can be estimated by (18.17): $U_* = Mg/K = \mu g T_*$. If $\mu \ll 1$, then the value of U_* is small.

The system (18.18) is singularly perturbed. The corresponding degenerate system is:

$$0 = -A\Pi \begin{bmatrix} \overline{u}_x \\ \overline{u}_y \end{bmatrix} + \begin{bmatrix} 0 \\ -1-h(\overline{y}-1) \end{bmatrix},$$
$$\frac{d}{dt}\begin{bmatrix} \overline{x} \\ \overline{y} \end{bmatrix} = \begin{bmatrix} 1 \\ -\frac{1}{\tau_0}(\overline{y}-1) \end{bmatrix}. \tag{18.19}$$

The last equation (18.19) can be written in the form:

$$\frac{d\overline{x}}{dt} = 1, \qquad \tau_0 \frac{d\overline{y}}{dt} + \overline{y} = 1, \tag{18.20}$$

which are the equations of motion along the servoconstraint. This system coincides with the initial equations of the servoconstraints (18.12) and (18.13) with accuracy up to insufficient scale multipliers.

Let us choose the matrix Π of the coordinate transformer, which realizes the servoconstraints. The uniform part of the adjoined system for (18.18) has the form:

$$\frac{d}{d\tau}\begin{bmatrix} u_x \\ u_y \end{bmatrix} = -A\Pi \begin{bmatrix} u_x \\ u_y \end{bmatrix}, \tag{18.21}$$

and in accordance with (18.8) it can be assumed that

$$\Pi = A^{-1}. \tag{18.22}$$

Then Eq. (18.21) takes the form:

$$\frac{d}{d\tau}\begin{bmatrix} u_x \\ u_y \end{bmatrix} + \begin{bmatrix} u_x \\ u_y \end{bmatrix} = 0.$$

For this equation the asymptotic stability of the trivial solution is obvious. For the realizability of (18.22) the matrix A should be nonsingular. It is seen from (18.16) that $|A| = 0$ when the columns A are proportional, that is, when the leg is straightened on the knee. This condition limits the length of the step.

Is it possible to avoid the necessity of calculating the matrix of the coordinate transformer according to (18.22), and is it possible to provide stability for the simpler variants of the transformer? Yes, it is. Let us choose, for example,

$$\Pi = \begin{bmatrix} 0 & 1 \\ -1 & 0 \end{bmatrix} H. \tag{18.23}$$

The characteristic polynomial of the system (18.21) for this case has the form $\lambda^2 + [(Y + X_1)H/\Delta]\lambda + H^2/\Delta$. The conditions of the asymptotic stability are:

$$Y + X_1 > 0,$$

$$\Delta = YX_1 - XY_1 > 0.$$

The first condition slightly limits the height of the trajectory; the second is satisfied while moving in the pose "the knee ahead."

For verifying these results, obtained in asymptotic approximation, a computer computation of the full equations (18.18) under the control (18.23) was performed. The magnitudes μ=0.1÷0.01 were chosen. The calculation revealed that after fading of fast motions in the boundary layer, the system with an error of order μ passes to the motion on servoconstraints, which is continued until the leg straightens at the knee. After this, the motion along the servoconstraints is collapsed.

The previously presented method of control was also applied for more complicated spatial motion of the two-legged and four-legged vehicles [35, 57, 58]. The stiff control on each support phase provides with high accuracy the motion according to the servoconstraints. The stable cyclic gait with changing support phases can be obtained by gluing degenerate trajectories for motion in accordance with servoconstraints.

While discussing the behavior of the systems — both for general form from Sec. 18.1 and connected with walking machines — the following doubt is often expressed. The real executive devices provide the control actions, limited in the absolute value. This circumstance is stipulated by the conditions analogous to (17.18). At the initial time moment the values of deviations u_s can occur as finite due to variance of the initial conditions and servoconstraints. The real executive devices with large coefficients K

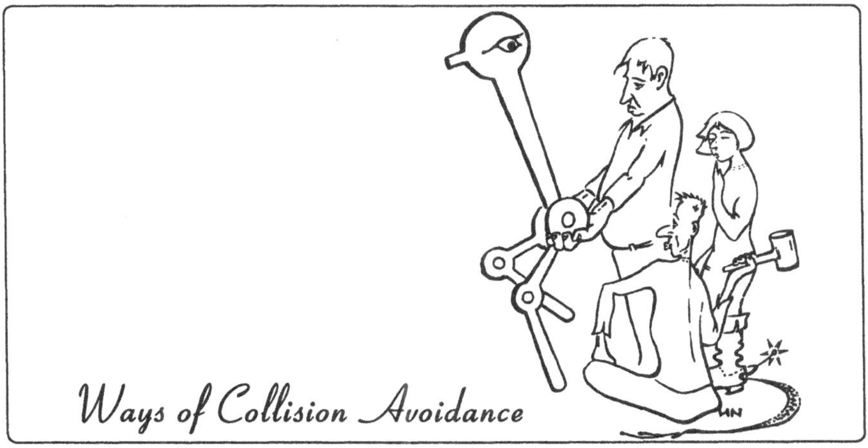

would attain the saturation interval f_s = const of their characteristics. The system would be nonstiff and all the previous conclusions would be not valid.

Let us consider possible methods of this collision avoidance.

1. The control of executive devices for saturation situation f_s = const is formed, which moves the system to the servoconstraints. Thus, for the problem of spacecraft motion control from Sec. 15.1, the choice $\tau_0 > 0$ guarantees the entrance to the servoconstraint $u = 0$ after the transient, developed during the limited time along the phase trajectory, combined with parabolic segments for $\operatorname{sgn} u = \pm 1$.

2. Let us provide an additional device which changes the initial conditions so that the conditions would fit the servoconstraint with a small error. For a legged vehicle it means that the vehicle gets a push and starts walking with speed $V_x(0) \approx V$.

3. For given initial conditions the servoconstraints are formed, providing the smallness of u_s during the motion process. For a legged vehicle with initial conditions $V_x(0) = 0$, $V_y(0) = 0$, the servoconstraints

$$U_x = V_x - V(1 - e^{-T/T_0}),$$

$$U_y = V_y + (Y - H)\,\frac{1 - e^{-T/T_0}}{T_0}$$

would satisfy this requirement instead of (18.13).

19 Precession and nutation models in gyro theory

19.1 Correctness conditions for an extended precession model

Usually gyroscopic systems consist of a lot of constructive elements; they are multiparametric and possess a great number of degrees of freedom. It is extremely hard to compose equations of motion in this case. That is why the simplified precession method for description of gyro systems has been used for some decades. It makes it possible to simulate slow components of gyro system motion with sufficient accuracy [15, 59].

In this section the correctness of the precession model is discussed and the model is extended to the class of gyroscopic systems with servoconstraints using stiff controls [43, 60 – 62].

Gyroscopic systems with the mechanical part composed of nondeformable elements connected with axial joints is considered. Some of these elements are rapidly rotating rotors of gyroscopes. The elements are enclosed by controlling loops; some of these loops provide hard control. The angles of the rotors' own rotation are assumed to be cyclic coordinates. Let n be the number of other elements. These elements are enumerated, and the trihedrons $x_{k1}x_{k2}x_{k3}$ are associated with them. The axis x_{k1} is directed along the axis of the joint connecting the kth element with one on which it is installed.

A set of mechanical systems is considered, consisting of the kth element and all elements and rotors installed on it. The first system in this set is the whole system installed on the base.

Let us compose the equations of motion of the system in a form of equations for its angular momenta projected onto axes x_{k1} of trihedrons $x_{k1}x_{k2}x_{k3}$ [15]:

$$
\begin{aligned}
&\frac{dG_{k1}}{dT} + \Omega_{k2}G_{k3} - \Omega_{k3}G_{k2} = L_k\left(K_s\Phi_s, K_\sigma\Phi_\sigma, \frac{T}{T_k}, \dots\right),\\
&G_{kp} = \sum_j \sum_{q=1}^{3} \Gamma_{kp}^{jq} I_{jq}\Omega_{jq} + \sum_m \Gamma_{kp}^{m1} H_m,\\
&\Omega_{jq} = \sum_{p=1}^{3} \Gamma_{jq}^{rp}\Omega_{rp} + \delta_{1q}\frac{d\Phi_j}{dT},\\
&\Gamma_{jq}^{rp} = \Gamma_{jq}^{rp}(\Phi_s, \Phi_\sigma),\\
&\delta_{11} = 1, \qquad \delta_{12} = \delta_{13} = 0; \qquad k = 1, \dots, n.
\end{aligned}
\tag{19.1}
$$

In (19.1) the following notations are used: G_{kp} are the components of the system angular momentum, Ω_{kp} are the components of the angular velocity of the kth trihedron projected onto its axes, L_k is the projection of the exterior with respect to the system moment onto the axis x_{k1}, Γ_{kp}^{jq} is the cosine of the angle between the pth axis of the kth trihedron and the qth axis of the jth trihedron, I_{jq} are the principal moments of inertia of the jth element (it is assumed for simplicity that axes of the jth trihedron coincide with principal axes of the jth element), H_m = const is the rotor angular momentum of the m-gyroscope, Φ_k is the turning angle of the kth element around its joint axis, K_s and K_σ are the coefficients characterizing effectiveness of control with respect to angles Φ_s and Φ_σ, and T_k is the characterizing time of dependence of L_k on explicit time. In (19.1) summation over m is that of serial numbers of gyroscopes, summation over j is that of serial numbers of the elements of the kth system under consideration, and r is the serial number of the element on which the jth element is installed. The system (19.1) is normalized in accordance with the expressions:

$$t = \frac{T}{T_*}, \qquad h_m = \frac{H_m}{H_*}, \quad g_{kp} = \frac{G_{kp}}{G_*}, \quad i_{kp} = \frac{I_{kp}}{I_*},$$

$$\omega_{kp} = \frac{\Omega_{kp}}{\Omega_*}, \quad \gamma_{kp}^{jq} = \frac{\Gamma_{kp}^{jq}}{\Gamma_*}, \quad \varphi_s = \frac{\Phi_s}{\Phi_{s*}}, \qquad \varphi_\sigma = \frac{\Phi_\sigma}{\Phi_{\sigma*}},$$

$$l_k = \frac{L_k}{L_*}, \qquad k_s = \frac{K_s}{K_{s*}}, \quad k_\sigma = \frac{K_\sigma}{K_{\sigma*}}.$$

Let $H_* = \max\{H_m\}$, $I_* = \max\{I_{kp}\}$. The characterizing value of G_* is chosen to be equal to H_* because the gyro system moment of momentum is determined by the "large" rotor angular momenta of gyroscopes included in it. Functional dependence of L_k on its variables should be written in such a way that variation of the quantities $K_s\Phi_s$, $K_\sigma\Phi_\sigma$, T/T_k by values of the order of unity implies variation of the function L_k by value of L_* order, and $K_{s*}\Phi_{s*} = K_{\sigma*}\Phi_{\sigma*}$.

It is assumed that the system is subjected to hard control $Ks* \gg K\sigma*$ with respect to some angular variables Φ_s; then $\Phi_{s*} \ll \Phi_{\sigma*}$.

The following class of motion is considered, which is of great importance for gyro theory. Let the system elements be turned in the space at great angles with "slow" angular velocities of the order of precession of a "large" H_* under the action of a moment of order L_*. For this class of motion it is assumed that $\Gamma_* = 1$; $\Phi_{\sigma*} = 1$ for the angle variables Φ_σ not involved in the hard control, and $\Omega_* = L_*/H_*$ for angular velocity. The characterizing time is $T_* = \Phi_{\sigma*}/\Omega_* = H_*/L_*$. From the condition $\Phi_{s*} \ll \Phi_{\sigma*}$ previously discussed, it follows that $\Phi_{s*} \ll 1$. Further, let $\Phi_{s*} = \varepsilon$ and ε be the small parameter of the problem. In gyro theory it is often assumed that $\varepsilon = 1/3440$. Then φ_s is the numerical measure of the angle Φ_s by angular minutes.

By the result of Sec. 18.1, the hard control with respect to the variable Φ_s must generate a servoconstraint. For Eqs. (19.1) only the simplest servoconstraints $u_s = \Phi_s = 0$ are considered, similar to interframe corrections which are typical for a lot of gyrosystems [34, 63, 55]. More complicated servoconstraints have been considered by Novozhilov [55]. After the preceding normalization Eqs. (19.1) take the form:

$$
\begin{aligned}
&\frac{dg_{k1}}{dt} + \omega_{k2} g_{k3} - \omega_{k3} g_{k2} = l_k(k_s\varphi_s, k_\sigma\varphi_\sigma, \nu_k t, \ldots), \\
&g_{kp} = \mu \sum_j \sum_{q=1}^{3} \gamma_{kp}^{jq} i_{jq} \omega_{jq} + \sum_m \gamma_{kp}^{m1} h_m, \\
&\gamma_{jq}^{kp} = \gamma_{jq}^{kp}(\varepsilon\varphi_s, \varphi_\sigma), \\
&\omega_{sq} = \sum_{p=1}^{3} \gamma_{sq}^{rp} \omega_{rp} + \varepsilon \delta_{1q} \frac{d\varphi_s}{dt}, \\
&\omega_{\sigma q} = \sum_{p=1}^{3} \gamma_{\sigma q}^{rp} \omega_{rp} + \delta_{1q} \frac{d\varphi_\sigma}{dt}; \\
&\mu = \frac{T_1}{T_*}, \qquad \nu_k = \frac{T_*}{T_k}.
\end{aligned}
\tag{19.2}
$$

By analogy with Sec. 4.1, the quantities $T_1 = I_*/H_*$ and $T_* = H_*/L_*$ are called nutation and precession time constants, respectively. Gyroscopic systems in which $T_1 \ll T_*$ are considered. Then $\mu \ll 1$. It is assumed, as previously, that perturbations are slow: $T_k \sim T_*$. Then $\nu_k \sim 1$.

Equations (19.2) contain small parameter ε characterizing the accuracy of stabilization with respect to angles Φ_s. In a wide class of gyroscopic systems the parameters μ and ε have magnitudes of the same order (in navigation gyro systems usually have $\mu \sim 10^{-5} \div 10^{-4}$, $\varepsilon \sim 10^{-4}$). Further, without limitation of generality, it is assumed that $\mu = \varepsilon$. This can be reached by finite change of characterizing factors.

Let us show that the system (19.2) is singularly perturbed. The variables g_{kp}, γ_{kp}^{jq} and ω_{jq} for $q \neq 1$ are excluded. The system obtained is composed of equations of the first order with respect to variables φ_j and ω_{jq} $(q = 1)$. It is resolved with respect to $d\omega_{jq}/dt$ and $d\varphi_j/dt$, assuming that conditions of solvability are satisfied. In the system obtained, all the derivatives $d\omega_{jq}/dt$, $d\varphi_s/dt$ are multiplied by small parameter μ, and the quantities $d\varphi_\sigma/dt$ are not multiplied by it. Therefore, the variables ω_{jq}, φ_s are fast, and φ_σ are slow. For the variables ω_{jq} this fact is determined by high values of nutation frequencies and high values of H_*; for φ_s, by high speed of control implied by high values of coefficients K_s.

Transformations of the equations into Tikhonov form are nondegenerate. Hence an asymptotic approximation of (19.2) can be constructed directly, without performing these transformations. The degenerate system for Eqs. (19.2) according to Tikhonov is obtained by setting $\mu = \varepsilon = 0$ in it:

$$\begin{gathered}\frac{d\overline{g}_{k1}}{dt} + \overline{\omega}_{k2}\overline{g}_{k3} - \overline{\omega}_{k3}\overline{g}_{k2} = l_k(k_s\overline{\varphi}_s, k_\sigma\overline{\varphi}_\sigma, \nu_k t, \ldots),\\ \overline{g}_{kp} = \sum_m \overline{\gamma}_{kp}^{m1} h_m, \quad \overline{\gamma}_{jq}^{kp} = \overline{\gamma}_{jq}^{kp}(0, \overline{\varphi}_\sigma),\\ \overline{\omega}_{sq} = \sum_{p=1}^{3} \overline{\gamma}_{sq}^{rp}\overline{\omega}_{rp}, \quad \overline{\omega}_{\sigma q} = \sum_{p=1}^{3} \overline{\gamma}_{\sigma q}^{rp}\overline{\omega}_{rp} + \delta_{1q}\frac{d\overline{\varphi}_\sigma}{dt}.\end{gathered} \tag{19.3}$$

First let us consider the system without hard control. In this case there is no index sequence s in (19.3), and (19.3) differs from (9.2) only by the second equation. According to this equation, the angular momentum of the kth system is formed of only the rotor angular momenta of gyroscopes. Hence, in this case Eqs. (19.3) determine the traditional model of gyroscopic system precession [15, 59].

Now let us assume that there is a hard control with respect to variables φ_s in the system. Then there are two points of difference between Eqs. (19.2) and (19.3): the first, angular momentum for (19.3) is composed by the precession method, and the second, additional conditions $\varphi_s = 0$ everywhere except l_k. Conditions $\varphi_s = 0$ are, as in Sec. 18.1, the equations of servoconstraints imposed on the system. The expressions l_k are "the reactions of servoconstraints."

Thus, the traditional definition of the precession model in gyro theory is extended to the class of gyroscopic systems with hard control.

Let us formulate the conditions for correctness of the limit model.

The adjoined system for (19.2) is constructed by excluding variables g_{kp}, substituting $t = \mu\tau$, and setting $\mu = 0$ in the obtained equations.

$$\sum_j \sum_{q=1}^{3} \gamma_{kp}^{jq} i_{jq} \frac{d\omega_{jq}}{d\tau} + \sum_m h_m \sum_j \frac{\partial \gamma_{kp}^{m1}}{\partial \varphi_j} \left(\omega_{j1} - \sum_{p=1}^{3} \gamma_{j1}^{rp} \omega_{rp} \right)$$
$$+ \omega_{k2} \sum_m \gamma_{k3}^{m1} h_m - \omega_{k3} \sum_m \gamma_{k2}^{m1} h_m = l_k(k_s \varphi_s, k_\sigma \varphi_\sigma, \nu_k t, \ldots),$$
$$\frac{d\varphi_s}{d\tau} = \omega_{s1} - \sum_{p=1}^{3} \gamma_{s1}^{rp} \omega_{rp}, \qquad \omega_{jq} = \sum_{p=1}^{3} \gamma_{jq}^{rp} \omega_{rp}, \quad (q \neq 1), \tag{19.4}$$
$$\gamma_{jq}^{kp} = \gamma_{jq}^{kp}(0, \varphi_\sigma).$$

The slow variables φ_σ, t are assumed to be parameters.

The system (19.4) coincides with the traditional form of nutation equations for gyro systems. The latter, as it is known, may be obtained by linearizing the initial equations in the proximity of "frozen" values of slow angles. That is why the dynamic system, defined by Eqs. (19.4) may be called the nutation model of a gyroscopic system.

The most important (according to Sec. 11.2) conditions for passage to (19.3) are:

- existence of an equilibrium position for Eqs. (19.4) with respect to ω_{jq} and φ_s; and
- asymptotic stability of these equilibrium positions.

Remark 1. Let us consider a variety of gyroscopic systems with the same kinematic schemes and the same sets of gyroscopes. Systems may differ by their high-frequency features: moments of inertia I_{jq} of their elements, finite stiffness of construction elements, dynamical characteristics of control channels, and the like. It is assumed that the characterizing values of time for all these high-frequency features are much less than the characterizing value of precession time. Then all these systems are defined by the same precession system of equations.

Remark 2. Usually high-frequency motion components are hardly describable, because they are defined by distributed stiffness of the system, interior medium friction, damping of a construction, and so on. So it is not always possible to analytically verify the conditions of asymptotic stability of fast motions. In this case it is sufficient to make sure of stability of these motions at least by results of a natural experiment.

Remark 3. Condition of asymptotic stability of fast motions is not necessary. Precession equations may be formed as well for gyroscopic systems with undamped fast components of motion. In this case variants of asymptotic methods [62] instead of the Tikhonov theorem should be used for motion decomposition.

In the following example extended definition of the precession model (19.3) is significant.

19.2 Precession model for a gyrotachometer

According to Merkin [59], "we have no right to compose precession equations for such a device." However, extended definition of precession equations makes it possible to do this.

A scheme of the device is pictured in Fig. 41. A gyroscope with rotor angular momentum H is enclosed in an instrument case which has one degree of freedom with respect to the base. The base is rotating in absolute space.

Let $O\xi\eta\zeta$ be the trihedron associated with the base, where $O\xi$ is the axis of rotation of the case, $Oxyz$ is the trihedron associated with the case, and α is the angle of a turn of the case relative to the base. It is assumed that the base affects the case with the moment $M_x = -K\alpha - R\dot{\alpha}$, $(K, R = \text{const})$. For real devices this action is always hard, so its coefficient K is large and, therefore, α is small.

Let us compose in accordance with (19.3) the precession equations of the device using the given dimensional notations of the problem. The equation for angular momentum of the device projected onto the axis x of the trihedron $Oxyz$ should be written. While calculating the vector of angular momentum $\vec{G}$, we take into account only the rotor angular momentum of gyro H, and servoconstraint $\alpha = 0$ is imposed on the system. Then $\vec{G} = \vec{\imath}\cdot 0 + \vec{\jmath}\cdot 0 + \vec{k}H$. The vector $\vec{\Omega}$ of angular velocity of the trihedron $Oxyz$ is represented by the expansion $\vec{\Omega} = \vec{\imath}\Omega_\xi + \vec{\jmath}\Omega_\eta + \vec{k}\Omega_\zeta$, Ω_ξ, Ω_η, Ω_ζ of the projection of angular velocity of the base onto the axes of the trihe-

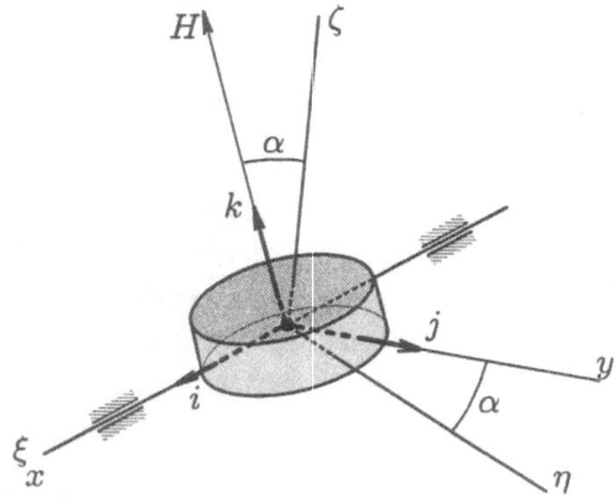

Fig. 41

dron associated with it. The result, according to (19.3), is the precession equation:

$$\Omega_\eta H = M_x, \tag{19.5}$$

where M_x is the reaction of the finite servoconstraint.

It is seen from (19.5) that measurement of the equilibrium value of an elastic moment makes it possible to measure the projection Ω_η of the base angular velocity. As usual, an elastic moment is implemented by an electromagnetic drive; therefore, we are to measure the current in the drive-feed circuit.

Let us verify the correctness of precession equation (19.5). Let us write down adjoined equations according to (19.4). They linearly depend on fast variables of the system and contain moments of inertia only as multipliers of the highest derivatives. The right-hand sides of the equations contain hard actions. According to (19.4), and keeping initial dimensional notations, we obtain:

$$I_x \frac{d}{dT}\Omega_x + H\Omega_\eta = -K\alpha - R(\Omega_x - \Omega_\xi), \qquad \Omega_x = \Omega_\xi + \frac{d\alpha}{dT}. \tag{19.6}$$

Assuming that motion of the base is slow compared to motion with respect to α, we accept that Ω_ξ, Ω_η in (19.6) are parameters. Then (19.5) takes the form

$$I_x \frac{d^2}{dT^2}\alpha = -K\alpha - R\frac{d}{dT}\alpha - H\Omega_\eta. \tag{19.7}$$

It is evident that equilibrium positions of the system (19.7) with respect to α and for (19.6) with respect to Ω_x, α do exist; they are isolated and asymptotically stable. Hence, the use of precession model (19.5) is correct.

In Merkin [59], a necessary condition for precession model validity consists of nonsingularity of a skew-symmetric matrix of system gyroscopic coefficients. A gyrotachometer possesses one "defining coordinate" and this condition does not hold for it. That is why Merkin [59] states the impossibility of a gyrotachometer precession model.

19.3 Precession model of a three-axis force gyrostabilizer

Let us sketch the main methodological tasks of this section.

Customary technique of mechanical systems analysis include the following steps.

1. Making assumptions forming a mechanical model of a system.
2. Constructing a mathematical model; that is, composing equations for description of the mechanical model.
3. Mathematical experiment; that is, analyzing the equations in some way.

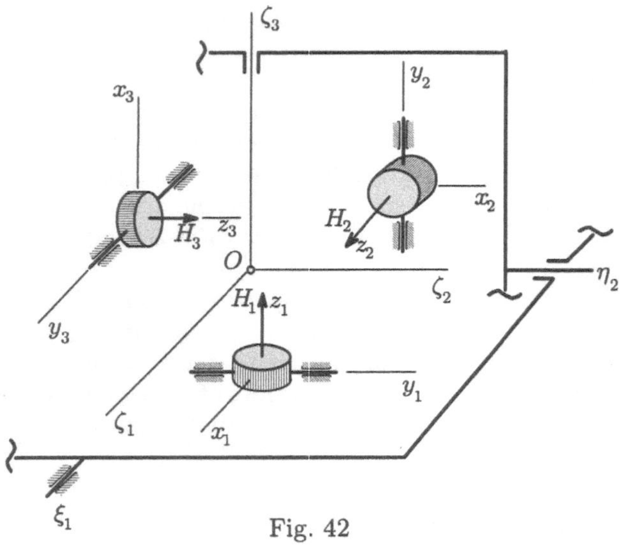

Fig. 42

The three-axis stabilizer is the first example in this book for which this program is, as it seems, quite impracticable. It has been estimated that complete equations for a gyrostabilizer would contain a few thousand summands. Obviously, there is no sense in obtaining such equations, even by computer.

So, such cumbersome objects are described by simplified mechanical models. There exists a variety of such models for various classes of object motion. The next step is composing equations for these simplified models corresponding to definite classes of motion. For gyroscopic systems such simplified models are the precession model with Eqs. (19.3) and nutation model with Eqs. (19.4).

Let us compose simplified equations for a gyrostabilizer, not even trying to write down the complete system of equations. This approach does not demand the performance of all steps of fractional analysis, that is, to normalizing equations, introducing small parameters, and so on. In this section the equations of limit models are composed using the structure (19.3) and (19.4) and initial notations of a problem.

A kinematic scheme of the device is shown in Fig. 42. The mechanical part of the system consists of a three-degree Cardano gimbal in which a stabilizing platform is installed. On this platform case axes of three gyroscopes Γ_1, Γ_2, Γ_3 are installed. Gyroscopes control the drive along axes 1, 2, 3 of the Cardano gimbal of the platform.

It is assumed that the system construction is absolutely rigid, the platform gimbal axes intersect at a point O, and the gyroscope gimbal axes intersect at points O_1, O_2, O_3.

Two individual features of a three-axis gyrostabilizer should be noted: it is composed of three one-axis stabilizers of the same type, and all three gyroscopes are installed on one element of the construction, on the platform. That is why it is convenient to renounce the general notation of Sec. 19 and introduce a new specific one. Let $OX_0Y_0Z_0$, $O\xi_1\xi_2\xi_3$, $O\eta_1\eta_2\eta_3$, $O\zeta_1\zeta_2\zeta_3$, $O_1x_1y_1z_1$, $O_2x_2y_2z_2$, $O_3x_3y_3z_3$ be right orthogonal trihedrons associated with the base, the exterior and interior rings of the Cardano gimbal platform, with the platform and the gyroscope cases, respectively. A position of the trihedrons $\xi_1\xi_2\xi_3$, $\eta_1\eta_2\eta_3$, $\zeta_1\zeta_2\zeta_3$ is obtained from $X_0Y_0Z_0$ as a result of turning at the angles α_1, α_2, α_3 around the successive positions of axes ξ_1, η_2, ζ_3 which coincide with the axes 1, 2, 3. A position of the trihedron $O_1x_1y_1z_1$ is obtained from $O\zeta_1\zeta_2\zeta_3$ by a shift of the coordinates origin from point O to point O_1, and a turn at angle β_1 around the case axis y_1 up to coincidence of the axis z_1 with the spin axis of the rotor of the gyroscope Γ_1. The position of trihedrons $O_2x_2y_2z_2$, $O_3x_3y_3z_3$ is defined quite similarly by angles β_2 and β_3. The remaining notation of Sec. 19.1 is preserved.

Assume that the rotors' angular momenta of gyroscopes are large, and the drive control with respect to angles β_1, β_2, β_3 is hard. Under these conditions the precession model of the gyrostabilizer may be constructed. It will be obtained if, in accordance with (19.3), the moment of momentum includes only rotor angular momenta of the gyroscopes, and servoconstraints $\beta_1 = \beta_2 = \beta_3 = 0$ are imposed on the system.

Because of the specific features of the system, it is more convenient to compose its precession equations using a set of mechanical subsystems that differs from the set considered in Sec. 19.1.

For gyroscopes Γ_1, Γ_2, Γ_3 by projection onto axes y_1, y_2, y_3 of trihedrons $O_1x_1y_1z_1$, $O_2x_2y_2z_2$, $O_3x_3y_3z_3$ it follows that:

$$\begin{aligned} -H_1\Omega_{x_1} &= M_1, & -H_2\Omega_{x_2} &= M_2, & -H_3\Omega_{x_3} &= M_3, \\ \Omega_{x_1} &= \Omega_{\zeta_1}, & \Omega_{x_2} &= \Omega_{\zeta_2}, & \Omega_{x_3} &= \Omega_{\zeta_3}, \end{aligned} \tag{19.8}$$

where M_1, M_2, M_3 are the moments applied to the gyroscopes along axes y_1, y_2, y_3. They consist of control moments and perturbation moments generating drift.

Equations for the platform with the gyroscopes installed on it are the following.

$$\begin{aligned} \Omega_{\zeta_2}H_1 - \Omega_{\zeta_3}H_3 &= M_{\zeta_1} + L_{\zeta_1}, \\ \Omega_{\zeta_3}H_2 - \Omega_{\zeta_1}H_1 &= M_{\zeta_2} + L_{\zeta_2}, \\ \Omega_{\zeta_1}H_3 - \Omega_{\zeta_2}H_2 &= M_{\zeta_3} + L_{\zeta_3}. \end{aligned} \tag{19.9}$$

Here M_{ζ_1}, M_{ζ_2}, M_{ζ_3} are the moments of the Cardano gimbal acting on the platform; L_{ζ_1}, L_{ζ_2}, L_{ζ_3} are the system perturbations moments reduced to the platform axes.

Let us write down three equations of statical equilibrium for inertialess Cardano gimbal rings: for both rings (in projection onto the axis ξ_1), for the interior one (in projection onto the axis η_2), and for the inertialess element connecting the platform and the motor along the axis ζ_3 (in projection onto this axis). The equations in matrix form are:

$$\begin{bmatrix} L_1 \\ L_2 \\ L_3 \end{bmatrix} + \Phi(\alpha_2, \alpha_3) \begin{bmatrix} -M_{\zeta_1} \\ -M_{\zeta_2} \\ -M_{\zeta_3} \end{bmatrix} = 0. \tag{19.10}$$

Here components of the first summand are control moments L_1, L_2, L_3 produced by the motor with respect to Cardano gimbal axes 1, 2, 3; components of the second summand are actions of the platform; square matrix $\Phi(\alpha_2, \alpha_3)$ is the direction cosine matrix of axes ξ_1, η_2, ζ_3 with respect to the axes of trihedron $\zeta_1\zeta_2\zeta_3$.

The preceding equations should be completed by the equations of control:

$$\begin{bmatrix} L_1 \\ L_2 \\ L_3 \end{bmatrix} = -K\Pi(\alpha_2, \alpha_3) \begin{bmatrix} \beta_1 \\ \beta_2 \\ \beta_3 \end{bmatrix}. \tag{19.11}$$

Here K is a large coefficient, and $\Pi(\alpha_2, \alpha_3)$ is the matrix of coordinates transform.

It is chosen that $\Pi = \Phi$; then from (19.10) and (19.11) under condition $|\Phi| \neq 0$ it follows that

$$\begin{bmatrix} M_{\zeta_1} \\ M_{\zeta_2} \\ M_{\zeta_3} \end{bmatrix} = -KE \begin{bmatrix} \beta_1 \\ \beta_2 \\ \beta_3 \end{bmatrix}.$$

In this case control of slow motion components is implemented independently for each one-axis channel of stabilization.

In practice, simplified versions of a coordinate transformator are usually used. Let us suppose, for example, that a gyrostabilizer is installed on the base which slightly perturbs the system with respect to the angles α_1, α_2. Then a coordinate transformator with matrix $\Pi = \Phi(0, \alpha_3)$ can be chosen.

Let us discuss the precession equations obtained. By (19.8),

$$\Omega_{\zeta_1} = -\frac{M_1}{H_1}, \qquad \Omega_{\zeta_2} = -\frac{M_2}{H_2}, \qquad \Omega_{\zeta_3} = -\frac{M_3}{H_3}. \tag{19.12}$$

Hence, turns of the platform in space are determined only by the moments applied to the gyroscopes.

The moments applied in this motion to the platform by the Cardano gimbal can be obtained from (19.9) and (19.12):

$$\begin{aligned} M_{\zeta_1} &= -L_{\zeta_1} - \frac{H_1}{H_2} M_2 + M_3, \\ M_{\zeta_2} &= -L_{\zeta_2} - \frac{H_2}{H_3} M_3 + M_1, \\ M_{\zeta_3} &= -L_{\zeta_3} - \frac{H_3}{H_1} M_1 + M_2. \end{aligned} \tag{19.13}$$

Motor moments, that is, reactions of servoconstraints, are determined by substitution of (19.13) into (19.10).

19.4 Two-step method for stability approval of the nutation model for a three-axis gyrostabilizer

In this section a nutation model of a gyrostabilizer [64 – 66] is examined. In order to avoid cumbersome calculations, a special case of a problem is considered in which a base is stationary, and nutation oscillations take place near position $\alpha_2 = \alpha_3 = 0$. In Bragin et al. [34] and Novozhilov [55] the problem of an indicator gyrostabilizer has been considered in a case of arbitrary evolutions of a base and great angles of turn of a Cardano gimbal.

According to (19.4), the nutation equations of gyrostabilizer motion along axis 1 have the form:

$$\begin{aligned} I_1 \frac{d}{dT} \Omega_{\zeta_1} + H_1 \Omega_{y_1} &= M_{\zeta_1} - R_1 \Omega_{\zeta_1}, \\ I_{y_1} \frac{d}{dT} \Omega_{y_1} - H_1 \Omega_{\zeta_1} = 0, \quad \Omega_{y_1} &= \Omega_{\zeta_2} + \frac{d\beta_1}{dT}, \\ \tau_1 \frac{dM_{\zeta_1}}{dT} + M_{\zeta_1} &= -K_1 \beta_1, \qquad (1\ 2\ 3) \end{aligned} \tag{19.14}$$

where I_1 is the total moment of inertia of all elements of the system taking part in motion with angular velocity Ω_{ζ_1} and I_{y_1} is the gyroscope moment of inertia with respect to its case axis. Dynamical properties of the stabilization channel are described by an element of the first order. Coefficient R_1 takes into account mechanical and electrical damping in axis 1. The symbolic expression (1 2 3) means that the equations corresponding to two other stabilization channels for the axes 2 and 3 are obtained from (19.14) by cyclic permutation of indices 1, 2, 3. Control coefficients K_1, K_2, K_3 corresponding to different channels are assumed to be, generally speaking, also different. In (19.14) the moments M_1, M_2, M_3 are assumed to be equal to zero. It means that this model is not valid for floated gyroscopes with high friction. Slowly varying summands of M_1, M_2, M_3 may be omitted, as they are not necessary for further analysis of stability.

After excluding the variables Ω_{y_1}, M_{ζ_1}, β_1 in (19.4), the system of equations with respect to unknowns Ω_{ζ_i} takes the form:

$$[(I_1D^2+R_1D+h_1)(\tau_1 D + 1)H_1D+h_1K_1]\,\Omega_{\zeta_1} - H_1K_1D\Omega_{\zeta_2} = 0$$
$$h_1 = H_1^2/I_{y_1}, \qquad D = d/dT. \qquad (1\ 2\ 3) \tag{19.15}$$

Let us consider the characteristic determinant

$$\Delta(\lambda) = |\,[M_{ij}]\,|, \qquad (i,j=1,2,3)$$

for the system of equations (19.15). Here λ is a complex variable and M_{ij} are the elements of the characteristic matrix:

$$M_{ii} = (I_i\lambda^2 + R_i\lambda + h_i)(\tau_i\lambda + 1)H_i\lambda + h_iK_i,$$

$$M_{12} = -H_1K_1\lambda, \quad M_{23} = -H_2K_2\lambda, \quad M_{31} = -H_3K_3\lambda,$$

$$M_{13} = M_{21} = M_{32} = 0.$$

The factor $H_1K_1\lambda$ is taken out of the first row of the determinant, the factor $H_2K_2\lambda$ is taken out of the second one, and $H_3K_3\lambda$ out of the third. Then

$$\Delta(\lambda) = (H_1H_2H_3K_1K_2K_3\lambda^3)\Delta_0, \tag{19.16}$$

where

$$\Delta_0 = \begin{vmatrix} X_1 & -1 & 0 \\ 0 & X_2 & -1 \\ -1 & 0 & X_3 \end{vmatrix} \tag{19.17}$$

is the polynomial of the auxiliary variables

$$X_i = \frac{(I_i\lambda^2 + R_i\lambda + K_i)(\tau_i\lambda + 1)H_i\lambda + h_iK_i}{H_iK_i\lambda}, \quad (i = 1,2,3). \tag{19.18}$$

In the following, a special case is considered for which all stabilization channels are identical. Then $X_i \equiv X$ and the determinant (19.17) is the polynomial of the only variable X:

$$\Delta_0 = X^3 - 1. \tag{19.19}$$

An analysis of stability is performed in two steps [64, 67]. The first one consists of finding roots $X^{(i)}$ of the polynomial (19.19). Then this polynomial is factored:

$$\Delta_0 = (X - X^{(1)})(X - X^{(2)})(X - X^{(3)}). \tag{19.20}$$

Substituting (19.20) into (19.16) produces

$$\Delta = (HK\lambda)^3(X - X^{(1)})(X - X^{(2)})(X - X^{(3)}). \tag{19.21}$$

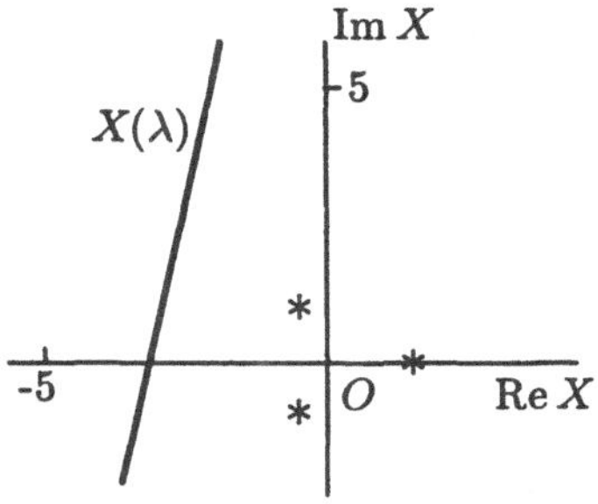

Fig. 43

At the second step analysis of the characteristic equation $\Delta = 0$ is separated into three independent problems consisting of separately analyzing each equation:

$$X - X^{(i)} = 0 \qquad (i = 1, 2, 3), \tag{19.22}$$

where dependence X on λ is given by (19.18).

Stability analysis of (19.22) may be based on different stability criteria. In Novozhilov [65] the left-hand side of the equation is represented as a polynomial of λ, and algebraic equations with unknown λ are constructed. Their coefficients are complex if roots $X^{(i)}$ are complex. Then stability conditions are formulated by means of the Hermite-Hurwitz generalized criterion.

In Novozhilov [66] conditions of stability are found by the argument principle. On a complex plane λ, a closed path C is considered consisting of two semicircles of the right half-plane with infinitesimal and infinitely large radii and the segments of imaginary axis closing the path. Then the stability condition is the following

$$\Delta_C \arg \Delta(\lambda) = 0, \tag{19.23}$$

where a variation of the argument of the function $\Delta(\lambda)$, as λ passes around the path C, is denoted by $\Delta_C \arg \Delta(\lambda)$. After substitution of (19.21) into (19.23) the stability conditions take the form:

$$\Delta_C \arg(X - X^{(i)}) = 0 \qquad (i = 1, 2, 3). \tag{19.24}$$

At first, a single one-axis stabilizer is considered. The conditions of its stability are obtained, if in (19.24) it is assumed that $X^{(i)} = 0$. In Novozhilov [66] a hodograph $X(\lambda)$ has been constructed for definite numerical values. Figure 43 shows its fragment in the vicinity of the zero point of a complex plane of the variable X. If the hodograph and the origin O are located as is shown in Fig. 43, then, by calculations, $\Delta_C \arg X$ is zero. This means that an independent one-axis gyrostabilizer is stable.

In the following, a three-axis gyrostabilizer is considered. Computations in accordance with (19.24) should be performed. If the points $X^{(i)}$ are located on the complex plane X on the same side of the hodograph as the point O, then the values of $\Delta_C \arg(X - X^{(i)})$ and $\Delta_C \arg(X - 0)$ are equal. Roots of the polynomial (19.19) are located on the unit circle. They are represented as stars in Fig. 43. It is clearly seen in Fig. 43 that each of the three roots satisfies the conditions (19.24).

In Bragin et al. [34] and Novozhilov [66] the frequency method previously described is extended to the case of different one-axis gyrostabilizers: $X_i \neq X$. Under such conditions stability analysis may be reduced to three separate problems determined by (19.22), where, by now, the $X^{(i)}$ are algebraic functions of the complex variable λ.

Thus, the expression (19.24) makes it possible to choose the parameter values guaranteeing stability of the system. Moreover, the conditions of the Tikhonov theorem providing the validity of the use of precession equations (19.8) through (19.11) are also satisfied.

Remark. The previously described two-step method for analysis of stability is successfully applicable to investigation multichannel control systems consisting of identical one-channel subsystems. So, this method has been used in the stability analysis of a two-channel control system of a spacecraft angular motion [68], three- and four channel gyroscopic systems [34, 55], the six-channel control system for the six-legged dynamical simulator [69], and the ten-channel control system of a magnetic train suspension [33].

Without separation into two steps, direct stability analysis using the initial form of the characteristic equation would be extremely cumbersome.

20 Mathematical model of a "man – artificial-kidney" system

Removing toxins, that is, metabolic products, from an organism can be carried out by an "artificial kidney" apparatus at acute or chronic renal insufficiency. The process of blood cleaning, called hemodialysis, is developed in the following way. Human blood taken, as usual, out of an artery, flows along a pipe into a dialyzer and then, after cleaning, is returned to a vein. The dialyzing solution balanced with blood (with respect to salt composition) moves through a dialyzer in opposite direction to blood flow. These two flows are separated by a semipervious membrane.

The problem of mathematical modeling of the hemodialysis process was raised in research carried out under the leadership of the outstanding physician, G.P.Kulakov (1926 – 1987). This problem is analyzed in this section.

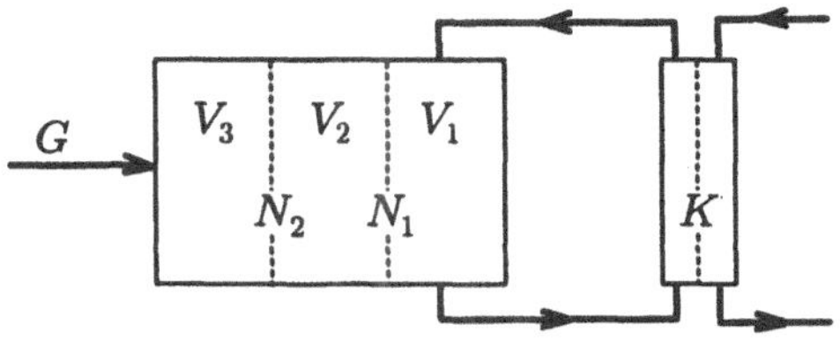

Fig. 44

A three-pool model of a "man-artificial-kidney" system (see, e.g., Lewis [70]) shown in Fig. 44 is considered. The interior liquid medium of an organism is characterized by three volumes V_1, V_2, V_3. Volume V_1 is the total of the blood circulatory system and V_2 and V_3 are volumes of intercellular and intracellular liquid, respectively. These volumes of liquid are separated by semipervious membranes through which the toxin diffusion process is developed. The varying of pool volumes caused by osmotic pressure is ignored. Toxin mass transfer does not change the volumes because toxin concentration in the solution is small. It is assumed that the velocity of mass transfer of a considered kind of toxin through the membranes due to diffusion is proportional with coefficients N_1, N_2 to the difference of this toxin concentration at both sides of the membrane.

Let X_1, X_2, X_3 be concentrations of a given toxin in corresponding pools, and G be the quality of the toxin produced by the organism during the time unit. It is supposed that the toxin is produced only in intercellular volume V_3. Let G_1 be a quantity of toxin removed from the organism per time unit by the dialyzer. It is determined by the expression $G_1 = A_1(X_1 - X_1^\alpha)$, where A_1 is the velocity of blood volume flow through the dialyzer; X_1 and X_1^α are the toxin concentrations in the blood at the input and output of the dialyzer, respectively. The expression for G_1 is transformed as follows.

$$G_1 = [A_1(X_1 - X_1^\alpha)/X_1]\,X_1 \equiv KX_1. \tag{20.1}$$

Here multiplier $K = A_1(X_1 - X_1^\alpha)/X_1$ is called the clearance of the dialyzer. The magnitude of K does not depend on the current concentration value and is equal to the volume of blood completely cleared by the dialyzer during the time unit. For the ideal dialyzer, while $X_1^\alpha = 0$, clearance is equal to blood flow velocity: $K = A_1$. Clearance dependence of blood flow and dialyzing solution velocities and membrane parameters has been determined by Novozhilov et al. [71]. A graph of the function $K = K(A_1)$ is the principal passport characteristic of a dialyzer. It is a result of experiments carried out by the apparatus designers.

While treating a patient by an artificial kidney apparatus, processes are being developed at a few hierarchical kinds of time scale.

The slowest ones, which last for a few years, are the processes of chronic illness evolution.

The processes of toxin accumulation during the period between hemodialysis procedures lasts for a few days. A person with complete renal insufficiency needs three such procedures a week to survive.

It takes five to seven hours to complete a single hemodialysis procedure. For this period toxin concentration in the blood decreases to an acceptable level.

The time period for complete liquid media circulation in a human organism varies from a few minutes to a few dozens of minutes. This is specified by high velocities of liquid volume flows in an organism. For example, the velocity of liquid flow through the heart is five liters per minute; that through the kidney is approximately four times less.

Elementary blood volume passes through a dialyzer for about one minute.

Now we analyze the dynamics of a single hemodialysis procedure. It is developed at an intermediate time scale among ones previously mentioned. Variables of all relatively slow processes are assumed to be "frozen": G, V_1, V_2, V_3, N_1, $N_2 = \text{const}$; it is assumed that all relatively fast processes are in their balanced states, concentrations X_1, X_2, X_3 are the same, and the dialyzer is regarded to be an inertialess element.

These a priori assumptions may be justified by the customary arguments of fractional analysis.

Let us compose equations of the system motion. Let V_1X_1, V_2X_2, V_3X_3 designate quantities of a toxin contained in each of the volumes. Rates of these quantities' change should be calculated. We should account for mass transfer from one pool to another, toxin production of the organism and its removal by an artificial kidney. In the extreme case, when the natural kidneys do not take part in toxin removal, the equations are

$$\begin{aligned}
&\frac{d}{dT}(V_1X_1) = -KX_1 + N_1(X_2 - X_1),\\
&\frac{d}{dT}(V_2X_2) = -N_1(X_2 - X_1) + N_2(X_3 - X_2),\\
&\frac{d}{dT}(V_3X_3) = -N_2(X_3 - X_2) + G.
\end{aligned} \tag{20.2}$$

Assuming that toxin distribution in the organism before the dialysis procedure is uniform, we may put

$$X_1(0) = X_2(0) = X_3(0) = X_0. \tag{20.3}$$

Let volume velocities of blood and dialyzing solution flows be constant. Then $K = \text{const}$.

Equations (20.2) describe the dynamics of the "Man – Artificial-kidney" system. Which magnitudes taking part in equations are known?

The clearance numerical value K is a documented characteristic of the dialyzer and is known with sufficiently high accuracy. Modern dialyzers have 200 cm^3/min and physiologically admissible $A_1 \approx 250$cm^3/min.

The estimates of some other parameter values are available. For example, the mean total quantity of liquid in organism $V = V_1 + V_2 + V_3$ is about 60% of body weight: $V \sim 40$ liters; blood plasma constitutes approximately 5% of the weight: $V_1 \sim 3-4$ liters, intercellular liquid and, intracellular one, 15% of the weight: $V_2 \sim 10$, and 40% of the weight: $V_3 \sim 25$ liters [72].

Magnitude G may be estimated by the following reasoning. Let T_{d} be dialysis duration. An apparatus with clearance K is capable of clearing the whole organism liquid volume for this time; hence,

$$KT_{\mathrm{d}} = V. \tag{20.4}$$

Let T_{b} be a time period between two dialysis sequences; $T_{\mathrm{b}} \approx 50$ hours. An organism produces toxin at the velocity $G = \mathrm{const}$; therefore, the quantity of toxin accumulated for the time period T_{b} is VX_0, where X_0 is the initial toxin concentration before the dialysis procedure defined by (20.3). Equate these quantities:

$$GT_{\mathrm{b}} = VX_0. \tag{20.5}$$

From (20.4) and (20.5) it follows that:

$$G = \frac{T_{\mathrm{d}}}{T_{\mathrm{b}}} KX_0. \tag{20.6}$$

The numerical values of N_1, N_2 in (20.2) are unknown.

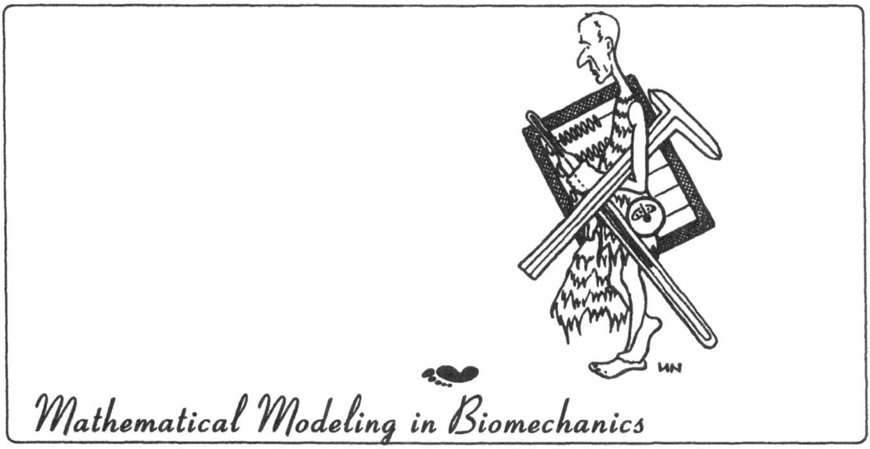

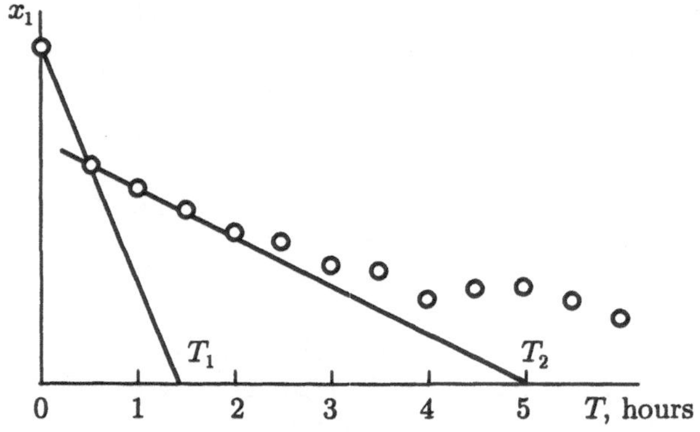

Fig. 45

Perform normalization in Eqs. (20.2): $x_1 = X_1/X_*$, $x_2 = X_2/X_*$, $x_3 = X_3/X_*$. Put $X_* = X_0$. Having divided each equation (20.2) by coefficient K, we convert the system (20.2) and (20.3) to the form:

$$\begin{aligned} T_1\frac{dx_1}{dT} &= -x_1 + n_1(x_2 - x_1), \\ T_2\frac{dx_2}{dT} &= -n_1(x_2 - x_1) + n_2(x_3 - x_2), \\ T_3\frac{dx_3}{dT} &= -n_2(x_3 - x_2) + g; \quad x_1(0) = x_2(0) = x_3(0) = 1, \end{aligned} \tag{20.7}$$

where $n_1 = N_1/K$, $n_2 = N_2/K$, $g = G/KX_0$. By (20.6), $g = T_\mathrm{d}/T_\mathrm{b}$. Here $T_1 = V_1/K, T_2 = V_2/K, T_3 = V_3/K$ designates partial time constants of the system. For parameter values previously estimated, $T_2 \sim 3T_1$, $T_3 \sim 5T_1$, $g < 1$.

Formulate an identification problem for parameters n_1, n_2, T_1, T_2, T_3, g of the system (20.7). Its data are specified by clinical experiment consisting of measuring toxin concentration $X_1(T)$ during a dialysis procedure for a fixed patient. The typical shape of the curve $x_1(T) = X_1/X_0$ is shown in Fig. 45.

Certain significant facts should be noted. The first is that the curve starts with steep decreasing, clearing as seen in Fig. 45. This leads to the supposition that there exists a boundary layer. The second is that the conditions of clinical experiment do not allow the performance of a great number of measurements. Usually blood is taken for analysis once an hour, sometimes once in a half hour. Therefore, the initial steep decreasing of the curve is specified by merely one or two experimental points. Customary identification methods are not suitable in such conditions; they require redundant data.

Let us try to separate the motions of the system (20.7) and to select equations of slow motions for their further identification. Being experienced in separating motions, we do not perform literally all the formal steps of fractional analysis. We may justify it by relative simplicity and visualizing of the system considered. Moreover, such an unusual approach will make it possible to demonstrate the significance of intermediate numerical estimations.

Being supplied with the experimental data shown in Fig. 45, let us estimate the numerical values of coefficients n_1, n_2. The following possible versions are to be analyzed.

(a). Let us assume that $n_1, n_2 \gg 1$. The system (20.7) may be regarded as "stiff" with respect to discrepancies $u_1 = x_2 - x_1$, $u_2 = x_3 - x_2$. In accordance with Sec. 18.1, equations approximating (20.7) may be obtained by imposing "servoconstraints" $u_1 = 0$, $u_2 = 0$. Addition of Eqs. (20.7) leads to the following system with respect to variables $x_1 = x_2 = x_3$.

$$\begin{aligned} T_0\frac{dx_1}{dT} &= -x_1 + g, \\ T_0 &= T_1 + T_2 + T_3 = V/K, \\ x_1(0) &= 1. \end{aligned} \tag{20.8}$$

Solution of these equations is the exponential function with time constant T_0. For parameter values given previously, $T_0 \sim 3$ hours.

Equation (20.8) describes a process taking place in a model with one pool, the volume of which is equal to the total liquid volume. This model does not explain the boundary layer effects that were discovered experimentally. So, we decline the assumption $n_1, n_2 \gg 1$.

(b). The next version is $n_1 \sim 1$, $n_2 \gg 1$. In this case the constraint $u_2 = x_3 - x_2 = 0$ is imposed on the system. Summation of the two latter equations of the limit model produces:

$$\begin{aligned} T_1\frac{dx_1}{dT} &= -x_1 + n_1(x_2 - x_1), \\ (T_2 + T_3)\frac{dx_2}{dT} &= -n_1(x_2 - x_1) + g. \end{aligned} \tag{20.9}$$

Given coefficient values specify $T_1/(T_2 + T_3) = V_1/(V_2 + V_3) \sim 0.1$. Under this condition (20.9) contains motion components with greatly divergent velocities.

Let us select in (20.9) slow motion components developing during a time interval sufficient for complete volume clearing. Let us normalize time $t = T/T_*$, having chosen $T_* = T_0$, where T_0 is the time constant of the model (20.8). Then (20.9) will be converted into

$$\begin{aligned} \tau_1 \frac{dx_1}{dt} &= -x_1 + n_1(x_2 - x_1), \\ (\tau_2 + \tau_3)\frac{dx_2}{dt} &= -n_1(x_2 - x_1) + g, \end{aligned} \tag{20.10}$$

where, by given parameter values $\tau_1 = T_1/T_0 \sim 0.1$, $\tau_2 + \tau_3 \sim 1$. In (20.10) τ_1 is assumed to be a small parameter of the problem.

Let us compose degenerate equations with respect to τ_1, corresponding to (20.10):

$$\begin{aligned} 0 &= -\overline{x}_1 + n_1(\overline{x}_2 - \overline{x}_1), \\ (\tau_2 + \tau_3)\frac{d\overline{x}_2}{dt} &= -n_1(\overline{x}_2 - \overline{x}_1) + g. \end{aligned}$$

Elimination of the variable $\overline{x}_2$ leads to

$$\begin{aligned} (\tau_2 + \tau_3)\frac{d\overline{x}_1}{dt} &= -\lambda_1 \overline{x}_1 + \lambda_1 g, \\ \lambda &= n_1/(1 + n_1). \end{aligned} \tag{20.11}$$

Let us estimate the evolution of the system (20.10) near the initial point and compare it with the experimental curve pictured in Fig. 45, beginning with the maximal velocity of x_1. The system (20.10) under conditions $x_1(0) = x_2(0) = 1$ implies

$$\left.\frac{dx_1}{dt}\right|_{t=0} = -\frac{1}{\tau_1}. \tag{20.12}$$

By (20.11) the maximal velocity near the initial point outside the boundary layer is estimated by the expressions:

$$\left.\frac{d\overline{x}_1}{dt}\right|_{t=0} = -\frac{\lambda_1(1-g)}{(\tau_2 + \tau_3)}. \tag{20.13}$$

From (20.12) and (20.13) the ratio of fast and slow motion component velocities is:

$$\left.\left[\frac{dx_1}{dt}\right] \Big/ \left[\frac{d\overline{x}_1}{dt}\right]\right|_{t\approx 0} = -\frac{(\tau_2 + \tau_3)}{\tau_1 \lambda_1 (1-g)} \sim 10. \tag{20.14}$$

Now let us evaluate this ratio, using experimental data. Maximal velocity inside the boundary layer may be estimated by plotting a tangent to the graph of Fig. 45 passing through the first two experimental points. The initial velocity of the slow components outside the boundary layer may be estimated by plotting the tangent through the second and third points. By visual observation, these points are located outside the boundary layer. The desired estimation is the ratio of segments OT_2 and OT_1 of the x-axis cut by the tangents. It is seen in the figure that

$$\left[\frac{dx_1}{dt}\right] \Big/ \left[\frac{d\overline{x}_1}{dt}\right] \Bigg|_{t\approx 0} \sim 3. \tag{20.15}$$

Discrepancy of estimations (20.14) and (20.15) means that assumption $n_2 \gg 1$ does not correspond to experimental results.

Moreover, having adopted this version, we discover a great discrepancy between preliminary estimation $T_1 = V_1/K \sim 15-20$ min and estimation of a smaller time constant obtained from the graph: $OT_1 \sim 90$ min. We ought to expect that identification of the parameter $T_1 \sim 15-20$ min performed for a given interval between measurements of 30 min is quite incorrect.

(c). The next version is $n_1 \gg 1$, $n_2 \sim 1$. As previously, the limit model is the result of imposing connections $u_1 = x_2 - x_1 = 0$ on (20.7). Summation of two fast equations of the limit model leads to

$$\begin{aligned}(T_1 + T_2)\frac{dx_2}{dT} &= -x_2 - n_2(x_2 - x_3), \\ T_3\frac{dx_3}{dT} &= n_2(x_2 - x_3) + g.\end{aligned} \tag{20.16}$$

This model is two-pool and, according to it, the ratio of slow and fast motion time constants $T_3/(T_1 + T_2) \sim 2 \div 3$ corresponds to the experimental estimation (20.15).

(d). The final version is $n_1, n_2 \sim 1$. In this case significant resistance of diffusion between the first and the second volumes ($n_1 \sim 1$) would cause boundary layer effects in time scale T_1, and the initial decreasing of the curve would be steeper than that shown in Fig. 45. This fact would violate the correspondence between experimental and prior estimations achieved in version (c). So, we decline this version.

Thus, the most suitable version, corresponding to experimental results, is the two-pool model specified by Eqs. (20.16).

These equations define the limit model of the hemodialysis process.

21 Approximate models of an aircraft motion

21.1 Models of zeroth approximation with respect to small parameters

It is well known that an aircraft motion is comprised of components that have a different hierarchy of time characteristics. For many years, different systems of equations for an aircraft-trajectory motion, for dynamic evolutions of mass center, and for angular motion with respect to the mass center have been used in computing practice (see, e.g.,Ostoslavsky [73], Byushgens and Studnev [74], and Bochkarev et al. [75]). For justification, qualitative considerations are commonly used, trustworthy but giving no possibility of estimating quantitative errors caused through oversight. In Novozhilov [4], Kuz'mak [76], and Borzov [77], methods of motion decomposition were used for the justification, but only an incomplete spectrum of motion components and some specific classes of flying vehicles were discussed. The following is an estimate of the possibile use of simplified models of an aircraft motion in a more complete formulation.

To reduce the tediousness of calculations, the following treatment is performed only for longitudinal aircraft motion. All the conclusions, without any limitations, are valid for an arbitrary motion in space.

Traditionally, equations for aircraft longitudinal motion dynamics are

$$\begin{gathered} M\frac{dV}{dT} = -Mg\sin\theta + P^T - \frac{1}{2}PV^2Sc_x, \\ MV\frac{d\theta}{dT} = -Mg\cos\theta + P^T\alpha + \frac{1}{2}PV^2Sc_y^\alpha, \\ I_{zz}\frac{d\Omega_z}{dT} = \frac{1}{2}PV^2Sb_a\left(m_z^\alpha + m_z^{\delta_z}\delta_z + m_z^{\omega_z}\frac{b_a}{V}\Omega_z + m_z^{\dot\alpha}\frac{b_a}{V}\frac{d\alpha}{dT}\right), \qquad (21.1) \\ \frac{d\vartheta}{dT} = \Omega_z, \qquad \vartheta = \theta + \alpha, \qquad P = P(Y), \\ \frac{dM}{dT} = -U, \quad \frac{dX}{dT} = V\cos\theta, \quad \frac{dY}{dT} = V\sin\theta, \quad \frac{dI_{zz}}{dT} = -W. \end{gathered}$$

Here M is an aircraft mass; X and Y are the coordinates of the mass center; V is its velocity; P^T is an engine thrust; P is an air density; θ, α, ϑ, and δ_z are angles of trajectory rise of attack, of pitch (see Fig. 46), and of elevator deflection; I_{zz}, S, and b_a are the moment of inertia, characteristic area, and length; W is a rate of moment of inertia change; U is a depletion of fuel per second; $c_x(\alpha)$, $c_y^\alpha\alpha$, and $m_z^\alpha + \ldots$ are expressions for aerodynamic characteristics (terms up to the second order are retained and all terms containing higher orders are dropped).

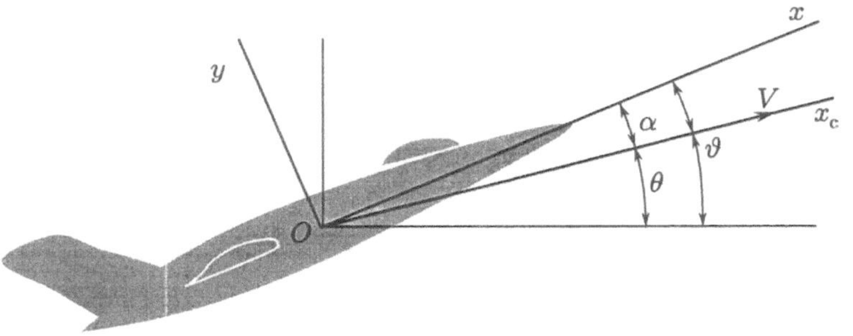

Fig. 46

Equations (21.1) are based upon the following assumptions: the Earth is flat and not rotating; an aircraft body configuration is fixed; the engine axis coincides with the longitudinal aircraft axis. The assumptions made are unimportant from the motion separation point of view, but simplify the calculations.

After eliminating the variable $\alpha = \vartheta - \theta$, in (21.1) and solving the system with respect to $d\Omega_z/dT$, $d\vartheta/dT$, $d\theta/dT$, the system of equations obtained should be normalized by introducing dimensionless analogues of quantities:

$$t = \frac{T}{T_*}, \quad m = \frac{M}{M_*}, \quad v = \frac{V}{V_*}, \quad p = \frac{P^T}{P^T_*},$$
$$\rho = \frac{P}{P_*}, \quad u = \frac{U}{U_*}, \quad w = \frac{W}{W_*}, \quad x = \frac{X}{X_*}, \tag{21.2}$$
$$y = \frac{Y}{Y_*}, \quad i_{zz} = \frac{I_{zz}}{I_*}, \quad \omega = \frac{\Omega_z}{\Omega_*}.$$

The angular variables θ, α, ϑ are assumed to be measured in radian measure and to take values of the order of unity. In this case there is no need to normalize θ, α, ϑ. Let us choose characterizing values for M_*, V_*, P^T_*, ... in accordance with (21.2). The values of M_*, V_*, U_*, W_*, I_*, X_*, Y_* are assumed to be equal to maximal ones for corresponding values for a specific aircraft type and a flight mode. It is also assumed that the characterizing values of aerodynamic forces and the thrust have the order of the aircraft weight. Then P_* and P^T_* can be found from $P_* V_*^2 S/2 = P^T_* = M_* g$. Here it is accepted that $X_* = Y_* = L_*$. The value of Ω_* is obtained from an estimate given by the simplified equation

$$I_* \frac{d^2\vartheta}{dT^2} = \frac{1}{2} P_* V_*^2 S b_a m_z^\alpha \vartheta.$$

With $m_z^\alpha \sim -1$, the time constant for this oscillating element can be estimated by the expression

$$T_1^2 = \frac{I_*}{P_* V_*^2 S b_a/2} = \frac{M_* r_*^2}{M_* g b_a} = \frac{r_*^2}{g b_a}, \tag{21.3}$$

where r_* is the central radius of inertia. Then if $\vartheta_* \sim 1$ is taken, $\Omega_* = 1/T_1$ may be accepted.

Dividing each equation that is normalized with respect to the (21.2) system by the combination of the characterizing values having the dimension of this equation yields

$$\begin{gathered}
\frac{T_2}{T_*} m \frac{dv}{dt} = -m \sin\theta + p - \rho v^2 c_x, \\
\frac{T_2}{T_*} m v \frac{d\theta}{dt} = -m \cos\theta + p(\vartheta - \theta) + \rho v^2 c_y^\alpha (\vartheta - \theta), \\
\frac{T_1}{T_*} i_{zz} \frac{d\omega_z}{dt} = \Big\{ m_z^\alpha (\vartheta - \theta) + m_z^{\delta_z} \delta_z + (m_z^{\omega_z} + m_z^{\dot\alpha}) \frac{T_0}{T_1} \frac{\omega_z}{v} \\
- m_z^{\dot\alpha} \frac{T_0}{T_2} \frac{1}{m v^2} \Big[- m \cos\theta + p(\vartheta - \theta) + \rho v^2 c_y^\alpha (\vartheta - \theta) \Big] \Big\} \rho v^2, \\
\frac{T_1}{T_*} \frac{d\vartheta}{dt} = \omega_z, \qquad \frac{T_3}{T_*} \frac{dm}{dt} = -u, \\
\lambda \frac{T_3}{T_*} \frac{dx}{dt} = v \cos\theta, \qquad \lambda \frac{T_3}{T_*} \frac{dy}{dt} = v \sin\theta, \\
\kappa \frac{T_3}{T_*} \frac{di_{zz}}{dt} = -w; \qquad \vartheta = \theta + \alpha, \qquad \rho = \rho(y).
\end{gathered} \tag{21.4}$$

Here $\lambda = (L_* U_*)/(V_* M_*) \sim 1$, $\kappa = (I_* U_*)/(W_* M_*) \sim 1$, and

$$T_0 = \frac{b_a}{V_*}, \qquad T_2 = \frac{V_*}{g}, \qquad T_3 = \frac{L_*}{V_*} \tag{21.5}$$

are partial time constants of the system.

In the table there are given orders of magnitude for partial time constants T_0, T_1, T_2, T_3 and their ratios for three aircraft types for subsonic, nearsonic and supersonic velocity, measured in meters per second. In estimates using (21.3) and (21.5) it was accepted that $b_a \sim 3$ m, $r_* \sim 3 \div 10$ m, and $L_* \sim 10^3 \div 10^4$ km. In the table, the maximal (from the considered intervals) values of μ and ε are given; that provide the worth estimates for the approximation errors.

V_*, m/s	100	300	1000
T_0, s	$3 \cdot 10^{-2}$	10^{-2}	$3 \cdot 10^{-3}$
T_1, s	$0.5 \div 2$	$0.5 \div 2$	$0.5 \div 2$
T_2, s	10	30	100
T_3, s	$10^4 \div 10^5$	$3(10^3 \div 10^4)$	$10^3 \div 10^4$
$\mu_1 = T_1/T_3$	$2 \cdot 10^{-4}$	10^{-3}	$2 \cdot 10^{-3}$
$\mu_2 = T_2/T_3$	10^{-3}	10^{-2}	10^{-1}
$\mu_3 = T_1/T_2$	$2 \cdot 10^{-1}$	10^{-1}	$2 \cdot 10^{-2}$
$\varepsilon_1 = T_0/T_1$	$6 \cdot 10^{-2}$	$2 \cdot 10^{-2}$	$6 \cdot 10^{-3}$
$\varepsilon_2 = T_0/T_2$	$3 \cdot 10^{-3}$	$3 \cdot 10^{-4}$	$3 \cdot 10^{-5}$

For the investigation we separate the slowest trajectory components of motion taking place on times $T \sim T_3$; then in (21.4) it is accepted that $T_* = T_3$. The system takes the form

$$\begin{aligned}
\mu_2 m\frac{dv}{dt} &= -m\sin\theta + p - \rho v^2 c_x,\\
\mu_2 mv\frac{d\theta}{dt} &= -m\cos\theta + p(\vartheta-\theta) + \rho v^2 c_y^\alpha(\vartheta-\theta),\\
\mu_1 i_{zz}\frac{d\omega_z}{dt} &= \Big\{m_z^\alpha(\vartheta-\theta) + m_z^{\delta_z}\delta_z + (m_z^{\omega_z} + m_z^{\dot\alpha})\varepsilon_1\frac{\omega_z}{v}\\
&\quad -m_z^{\dot\alpha}\varepsilon_2\frac{1}{mv^2}\Big[-m\cos\theta + p(\vartheta-\theta) + \rho v^2 c_y^\alpha(\vartheta-\theta)\Big]\Big\}\rho v^2,\\
\mu_1\frac{d\vartheta}{dt} &= \omega_z,\quad \frac{dm}{dt} = -u,\quad \lambda\frac{dx}{dt} = v\cos\theta,\quad \lambda\frac{dy}{dt} = v\sin\theta,\\
\kappa\frac{di_{zz}}{dt} &= -w;\qquad \vartheta = \theta+\alpha,\qquad \rho = \rho(y).
\end{aligned} \tag{21.6}$$

The system (21.6) is singularly perturbed with respect to small parameters μ_1, $\mu_2 \ll 1$. An unperturbed degenerate system is obtained from (21.6) by substituting $\mu_1 = 0$, $\mu_2 = 0$.

$$\begin{aligned}
0 &= -m\sin\theta + p - \rho v^2 c_x,\\
0 &= -m\cos\theta + p(\vartheta-\theta) + \rho v^2 c_y^\alpha(\vartheta-\theta),\\
0 &= m_z^\alpha(\vartheta-\theta) + m_z^{\delta_z}\delta_z,\qquad \omega_z = 0,\\
\frac{dm}{dt} &= -u,\quad \lambda\frac{dx}{dt} = v\cos\theta,\quad \lambda\frac{dy}{dt} = v\sin\theta,\quad \vartheta = \theta+\alpha.
\end{aligned} \tag{21.7}$$

In the dimensional notation according to (21.2), the system (21.7) coincides with the traditional system of equations, describing quasistatic aircraft trajectory motions.

In accordance with the Tikhonov – Vasil'eva theorem from Sec. 13 an error of the approximate solution (21.7) on time intervals of order $T_* = T_3$ will have an order of μ_1 for singularity in the equations describing angular motions, and have an order of μ_2 for singularity in the equations describing center mass motion. From the table it is easily seen that in the first case an error will have the order of fractions of a percent for all aircraft classes, but in the second, an error can attain to tens of percents for supersonic aircrafts ($V_* = 1000\ m/s$) having relatively small flight ranges ($L_* = 10^3$ km). It should be noted that the terms with factors ε_1, ε_2 from (21.6) in (21.7) vanish despite the magnitude of these factors.

The main condition for ensuring the validity of the preceding error evaluations is the asymptotic stability of system motion with respect to fast components developing in the time scales T_1, T_2. This condition is satisfied for any real flight because an aircraft is always controlled by an autopilot or a pilot.

While separating the dynamic components of a centermass motion taking place in time intervals of the order of T_2, let us assume $T_* = T_2$ in (21.4). The system (21.4) takes the form

$$
\begin{aligned}
m\frac{dv}{dt} &= -m\sin\theta + p - \rho v^2 c_x,\\
mv\frac{d\theta}{dt} &= -m\cos\theta + p(\vartheta - \theta) + \rho v^2 c_y^\alpha(\vartheta - \theta),\\
\mu_3 i_{zz}\frac{d\omega_z}{dt} &= \Big\{ m_z^\alpha(\vartheta - \theta) + m_z^{\delta_z}\delta_z + (m_z^{\omega_z} + m_z^{\dot\alpha})\varepsilon_1\frac{\omega_z}{v}\\
&\quad - m_z^{\dot\alpha}\varepsilon_2\frac{1}{mv^2}\Big[- m\cos\theta + p(\vartheta-\theta) + \rho v^2 c_y^\alpha(\vartheta - \theta)\Big]\Big\}\rho v^2,\\
\mu_3\frac{d\vartheta}{dt} &= \omega_z, \qquad \frac{dm}{dt} = -\mu_2 u,\\
\lambda\frac{dx}{dt} &= \mu_2 v\cos\theta, \qquad \lambda\frac{dy}{dt} = \mu_2 v\sin\theta,\\
\kappa\frac{di_{zz}}{dt} &= -\mu_2 w; \qquad \vartheta = \theta + \alpha, \qquad \rho = \rho(y).
\end{aligned}
\tag{21.8}
$$

The system (21.8) is regularly perturbed with respect to small parameters μ_2, ε_2 and singularly perturbed with respect to μ_3. Thr degenerate system for (21.8) with respect to these parameters can be written in the form

$$
\begin{aligned}
m\frac{dv}{dt} &= -m\sin\theta + p - \rho v^2 c_x, \\
mv\frac{d\theta}{dt} &= -m\cos\theta + p(\vartheta - \theta) + \rho v^2 c_y^\alpha(\vartheta - \theta), \\
\frac{dm}{dt} = 0, \quad \lambda\frac{dx}{dt} &= 0, \quad \lambda\frac{dy}{dt} = 0, \quad \frac{di_{zz}}{dt} = 0, \qquad (21.9)\\
0 &= m_z^\alpha(\vartheta - \theta) + m_z^{\delta_z}\delta_z, \\
\omega_z = 0; \quad \vartheta &= \theta + \alpha, \quad \rho = \rho(y).
\end{aligned}
$$

The system (21.9) in initial dimensional variables is traditionally used to describe fugoid oscillations, aerobatics, loops and the like.

Let us evaluate the errors caused by the transition from (21.8) to (21.9) in accordance with the table. By the Tikhonov – Vasil'eva theorem the error caused by neglecting components with μ_3 may, in general, run into values of some percents for supersonic aircrafts and values of orders of tens of percents for subsonic and nearsonic aircrafts.

Regular perturbations in (21.8) with respect to μ_2 and ε_2 according to the Poincaré theorem on time intervals of the order of T_2 cause errors not exceeding values of some percent order for all classes of flying vehicles mentioned in the table except for supersonic aircrafts with a small flying range, for which $\mu_2 = T_2/T_3 \sim 10^{-1}$. As before, the terms with the factor ε_1 from (21.8) in (21.9) vanish because $\omega_z = 0$ for all values of this parameter.

In order to separate high frequency components of the motion about a mass center on characteristic times of the order of T_1, we assume that

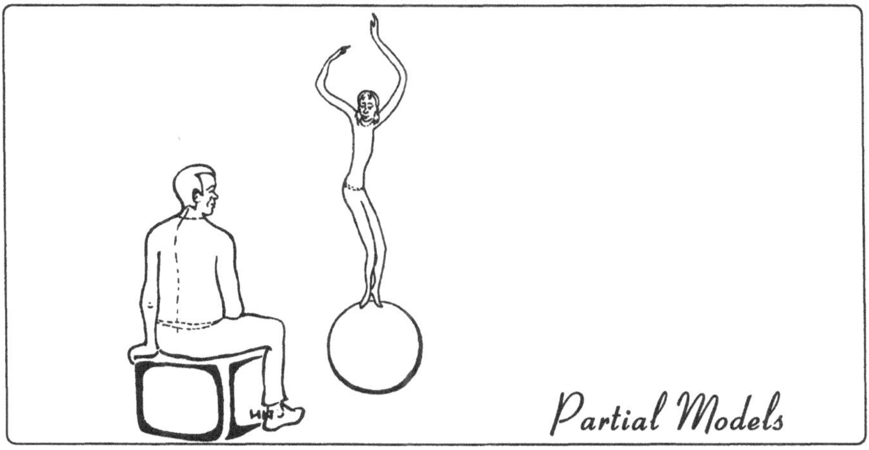

$T_* = T_1$ in (21.4). Then the system (21.4) takes the form

$$
\begin{aligned}
m\frac{dv}{dt} &= \mu_3(-m\sin\theta + p - \rho v^2 c_x),\\
mv\frac{d\theta}{dt} &= \mu_3[-m\cos\theta + p(\vartheta-\theta) + \rho v^2 c_y^\alpha(\vartheta-\theta)],\\
i_{zz}\frac{d\omega_z}{dt} &= \Big\{ m_z^\alpha(\vartheta-\theta) + m_z^{\delta_z}\delta_z + (m_z^{\omega_z} + m_z^{\dot\alpha})\varepsilon_1\frac{\omega_z}{v}\\
&\quad - m_z^{\dot\alpha}\frac{\varepsilon_2}{mv^2}\Big[-m\cos\theta + p(\vartheta-\theta) + \rho v^2 c_y^\alpha(\vartheta-\theta)\Big]\Big\}\rho v^2,\\
\frac{dm}{dt} &= -\mu_1 u, \qquad \lambda\frac{dx}{dt} = \mu_1 v\cos\theta,\\
\lambda\frac{dy}{dt} &= \mu_1 v\sin\theta, \qquad \kappa\frac{di_{zz}}{dt} = -\mu_1 w,\\
\frac{d\vartheta}{dt} &= \omega_z, \qquad \vartheta = \theta + \alpha, \qquad \rho = \rho(y).
\end{aligned}
\tag{21.10}
$$

The system (21.10) is regularly perturbed with respect to μ_1, μ_3, ε_1, ε_2. A degenerate system with respect to these small parameters is:

$$
\begin{aligned}
&\frac{dv}{dt} = 0, \quad \frac{d\theta}{dt} = 0,\\
&i_{zz}\frac{d\omega_z}{dt} = \left[m_z^\alpha(\vartheta-\theta) + m_z^{\delta_z}\delta_z\right]\rho v^2,\\
&\frac{d\vartheta}{dt} = \omega_z, \quad \vartheta = \theta + \alpha, \quad \frac{dm}{dt} = 0,\\
&\lambda\frac{dx}{dt} = 0, \quad \lambda\frac{dy}{dt} = 0,\\
&\frac{di_{zz}}{dt} = 0, \quad \rho = \rho(y).
\end{aligned}
\tag{21.11}
$$

According to this system slow variables m, x, y, I_{zz}, ρ can be accepted to be constants with the error of order μ_1, that is, small for all aircraft classes. Slow variables v, θ can be accepted to be constant with the error μ_3 having the order of some percent for supersonic and tens of percents for subsonic and nearsonic aircrafts.

A normalization of aircraft equations and introduction of small parameters for different classes of motion make it possible to formalize the composition of corresponding approximate equations, that is, approximate mathematical models of motion. Such approximate models for separated components of motion are called for brevity "partial" models.

21.2 Refined models of motion

The error estimates of partial models, previously carried out in a number of cases run into values of an order of tens of percents. Undoubtably these estimates can be improved if the initial data for an aircraft and a separate motion mode were more accurate. If required, the approximate motion models can be refined by constructing next approximations with respect to small parameters of the problem.

For example, let us construct a refined model of the center of mass dynamic evolution for an aircraft flying with nearsonic velocity $V_* = 300m/s$.

For this motion class according to the table from Sec. 21.1 in Eqs. (21.8),

$$\begin{gathered} \mu_1 = 10^{-3}, \qquad \mu_2 = 10^{-2}, \qquad \mu_3 = 10^{-1}, \\ \varepsilon_1 = 2 \cdot 10^{-2}, \qquad \varepsilon_2 = 10^{-4}. \end{gathered} \tag{21.12}$$

Equations (21.9) of the limit model are obtained from (21.8) at null values of all parameters from (21.12).

Let us construct the next approximation for (21.8) with respect to the largest small parameter $\mu_3 = 10^{-1}$ for slow motion components outside a boundary layer using the algorithm from Sec. 13.

$$\overline{v}_{(1)} = \overline{v}^{(0)} + \mu_3 \overline{v}^{(1)}, \quad \overline{\theta}_{(1)} = \overline{\theta}^{(0)} + \mu_3 \overline{\theta}^{(1)}, \ldots \tag{21.13}$$

The parameters μ_1, μ_2, ε_2 from (21.12) are supposed to be the values of the second and higher order, infinitesimal with respect to μ_3. The factor $\varepsilon_1(m_z^{\omega_z} + m_z^{\dot{\alpha}})$ in (21.8) is assumed to be a value of the first order with respect to μ_3 because for the real aircraft characteristics $m_z^{\omega_z}$, $m_z^{\dot{\alpha}} \sim 10$. The coefficients $m_z^{\omega_z}$, $m_z^{\dot{\alpha}}$ can be normalized also, $m_z^{\omega_z} = m_* \overline{m}_z^{\omega_z}, \ldots$, so that $\overline{m}_z^{\omega_z}$, $\overline{m}_z^{\dot{\alpha}} \sim 1$, and the role of a small parameter in (21.8) would be performed by the product $\varepsilon_1 m_* \sim \mu_3 = 10^{-1}$.

Then when solving the problem (21.13) in Eqs. (21.8) we can omit the terms containing all small parameters except μ_3 and ε_1:

$$\begin{aligned} m\frac{dv}{dt} &= -m\sin\theta + p - \rho v^2 c_x(\alpha), \\ mv\frac{d\theta}{dt} &= -m\cos\theta + (p + \rho v^2 c_y^{\alpha})\alpha, \\ \mu_3 i_{zz}\frac{d\omega_z}{dt} &= \left[m_z^{\alpha}\alpha + m_z^{\delta_z}\delta_z + \mu_3(m_z^{\omega_z} + m_z^{\dot{\alpha}})\frac{\varepsilon_1}{\mu_3}\frac{\omega_z}{v}\right]\rho v^2, \\ \mu_3\frac{d\vartheta}{dt} &= \omega_z, \quad \vartheta = \theta + \alpha. \end{aligned} \tag{21.14}$$

Here according to (21.9) we assume m, ρ, $i_{zz} = \text{const}$ and for definiteness it is accepted that $p = \text{const}$, $\delta_z = \delta_z(t)$.

Let us write down equations of zeroth approximation for (21.14) with respect to μ_3 outside a boundary layer:

$$
\begin{aligned}
m\frac{d\overline{v}^{(0)}}{dt} &= -m\sin\overline{\theta}^{(0)} + p - \rho\overline{v}^{(0)2}c_x,\\
m\overline{v}^{(0)}\frac{d\overline{\theta}^{(0)}}{dt} &= -m\cos\overline{\theta}^{(0)} + (p+\rho\overline{v}^{(0)2}c_y^\alpha)\overline{\alpha}^{(0)},\\
0 &= \left[m_z^\alpha\overline{\alpha}^{(0)} + m_z^{\delta_z}\delta_z\right]\rho\overline{v}^{(0)2},\\
0 &= \overline{\omega}_z^{(0)}, \quad \overline{\vartheta}^{(0)} = \overline{\theta}^{(0)} + \overline{\alpha}^{(0)}.
\end{aligned}
\tag{21.15}
$$

The equations of the next approximation with respect to μ_3 according to (13.16) are

$$
\begin{aligned}
m\frac{d\overline{v}^{(1)}}{dt} &= -m\sin\overline{\theta}^{(0)}\overline{\theta}^{(1)} - 2\overline{v}^{(0)}\overline{v}^{(1)}c_x(\overline{\alpha}^{(0)}) - \overline{v}^{(0)2}\left[\frac{\partial c_x}{\partial\alpha}\right]^{(0)}\overline{\alpha}^{(1)},\\
m\left(\overline{v}^{(0)}\frac{d\overline{\theta}^{(1)}}{dt} + \overline{v}^{(1)}\frac{d\overline{\theta}^{(0)}}{dt}\right) &= m\sin\overline{\theta}^{(0)}\overline{\theta}^{(1)}\\
&\quad +(p+\rho\overline{v}^{(0)2}c_y^\alpha)\overline{\alpha}^{(1)} + 2\overline{v}^{(0)}\overline{v}^{(1)}c_y^\alpha\overline{\alpha}^{(0)},\\
i_{zz}\frac{d\overline{\omega}_z^{(0)}}{dt} &= \left[m_z^\alpha\overline{\alpha}^{(1)} + (m_z^{\omega_z} + m_z^{\dot\alpha})\frac{\varepsilon_1}{\mu_3}\frac{\overline{\omega}_z^{(0)}}{\overline{v}^{(0)}}\right]\rho\overline{v}^{(0)2}\\
&\quad +2\overline{v}^{(0)}\overline{v}^{(1)}\left[m_z^\alpha\overline{\alpha}^{(0)} + m_z^{\delta_z}\delta_z\right],\\
\frac{d\overline{\vartheta}^{(0)}}{dt} &= \overline{\omega}_z^{(1)}, \qquad \overline{\vartheta}^{(1)} = \overline{\theta}^{(1)} + \overline{\alpha}^{(1)}.
\end{aligned}
\tag{21.16}
$$

Thus, to construct the approximate solution of (21.13), the following calculations should be carried out: for given initial conditions and predetermined control $\delta_z(t)$ the solution of the system (21.15) should be found; the solution obtained should be substituted into (21.16), and this system should be integrated with initial conditions being set according to (13.21). Obtained solutions for the systems (21.15) and (21.16) should be substituted into (21.13).

It is readily seen that obtaining the approximate solution (21.13) is by far a more laborious multistage task than integration of the initial system (21.14). For "large" values of the small parameter $\mu_3 \sim 10^{-1}$ the difficulties of computer integration of the stiff system (21.14) are not great. The approximate procedure seems to have no sense. An attempt to simplify calculations of a refined model is presented in the following.

It should be noted that the left-hand sides of the third equation in both systems (21.15) and (21.16) are equal to zero. For (21.16) this holds true due to the fourth equation of (21.15). Thus, if in the initial Eqs. (21.14) the left-hand side of the third equation is set to be equal to zero, then this will have no effect on the calculations of the first two members of the asymptotic expansion outside the boundary layer. The simplified system using new variables $\overline{v}$, $\overline{\theta}$ ... has the form:

$$\begin{aligned} m\frac{d\overline{v}}{dt} &= -m\sin\overline{\theta} + p - \rho\overline{v}^2 c_x, \\ m\overline{v}\frac{d\overline{\theta}}{dt} &= -m\cos\overline{\theta} + (p + \rho\overline{v}^2 c_y^\alpha)\,\overline{\alpha}, \\ 0 &= \left[m_z^\alpha\overline{\alpha} + m_z^{\delta_z}\delta_z + (m_z^{\omega_z} + m_z^{\dot\alpha})\varepsilon_1\frac{\overline{\omega}_z}{\overline{v}}\right], \\ \mu_3\frac{d\overline{\vartheta}}{dt} &= \overline{\omega}_z, \qquad \overline{\vartheta} = \overline{\theta} + \overline{\alpha}. \end{aligned} \tag{21.17}$$

This system is closed. This fact precludes the need to complete cumbersome two-step calculations according to (21.13), (21.15), and (21.16). The system of Eqs. (21.17) is simpler than the initial (21.14), and more accurate than equations of the zeroth approximation (21.15).

It is possible to compose Eqs. (21.17) using another technique: multiplying all equations of the system (21.16) by μ_3 and adding to them corresponding equations from (21.15). The obtained system of equations reduces to an identity on (21.13) (for terms of zeroth and first order in powers of μ_3 separately). With the small changes having an order of μ_3^2, the system obtained is reduced to the closed system of equations of the form (21.17) with respect to variables $\overline{v} \approx \overline{v}^{(0)} + \mu_3\overline{v}^{(1)}$, $\overline{\theta} \approx \overline{\theta}^{(0)} + \mu_3\overline{\theta}^{(1)}, \dots$.

To estimate asymptotic error of the solution for the system (21.17) the expressions $\overline{v} = \overline{v}^{(0)} + \mu_3\overline{v}^{(1)}$, $\overline{\theta} = \overline{\theta}^{(0)} + \mu_3\overline{\theta}^{(1)}, \dots$ should be substituted into it. The result of the substitution is not identically equal to zero and is determined by the second and higher order infinitesimal members with respect to μ_3. In this case the asymptotic error of the solution approximation for the initial system (21.14) outside the boundary layer by the system (21.17) coincides with (21.13).

Test computations show that approximation accuracy also substantially increases for fixed considerably large small parameter values. For test computations there were taken: $m = 1$, $i_{zz} = 1$, $c_x = 0.03$, $c_y^\alpha = 8$, $m_z^\alpha = -1.85$, $m_z^{\delta_z} = -1.6$, $m_z^{\omega_z} + m_z^{\dot\alpha} = -25$, $\varepsilon_1 = 0.02$, $\mu_3 = 0.2$. Analogous results were obtained for greater values of $\mu_3 = 0.4, 0.5, \dots$. The magnitudes $p = p_0 = \text{const}$, $\vartheta = \theta_0 = \text{const}$, $\delta_z = \delta_0 = \text{const}$ were determined by the static relations $p_0 = v_0^2 c_x$, $\vartheta_0 = (p_0 + \vartheta_0^2 c_4^\alpha)^{-1}$, $\delta_0 = -m_z^\alpha\vartheta_0/m_z^{\delta_z}$ for stable horizontal flight with $\theta = 0$, $v = v_0 = 1.5$. Harmonic variation of the elevator position was given as $\delta_z = \delta_0(1 + 0.1\sin t)$. The initial values were taken to be equal to their values for unperturbed flight: $\theta(0) = 0$, $v(0) = v_0$, $\vartheta(0) = \vartheta_0$, $\omega_z(0) = 0$. For these initial conditions with respect to fast variables the corrections of the form (13.22) are equal to zero.

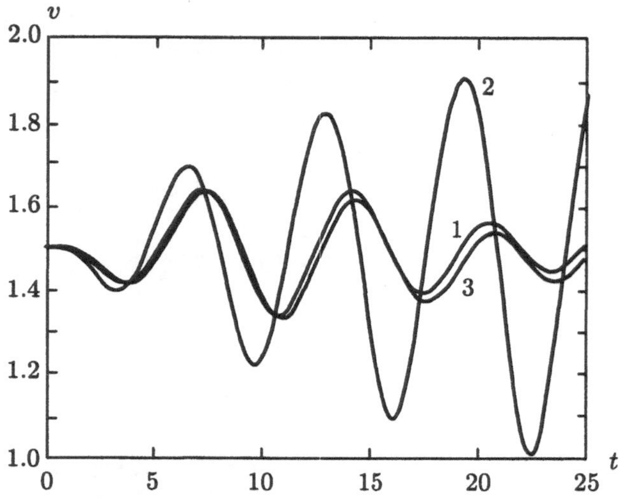

Fig. 47

In Fig. 47 graphs of v-dependence on time are shown. The solution of the system (21.14) is designated by number 1; solutions of the systems (21.15) and (21.17) are designated by numbers 2 and 3.

Let us explain the result obtained by a simple example. An equation for a linear oscillator with an harmonic perturbation is considered:

$$(T_1^2 D^2 + 2\zeta T_1 D + 1)x = f_0 \sin T/T_2.$$

Here $D = d/dt$ and the perturbation frequency is $1/T_2$. Let the perturbation be slow: $T_1 \ll T_2$. Let us construct amplitude and phase frequency characteristics for this system. It can be easily proved that for $\mu = T_1/T_2 \ll \zeta \sim 1$ the oscillator equation considered can be substituted by the equation of the first order

$$(2\zeta T_1 D + 1)x = f_0 \sin T/T_2.$$

The error caused by this simplification with respect to amplitude and phase of forced oscillations is a value of the order μ^2.

Thus, it is possible to construct refined models similar to (21.17) if partial components of the motion with respect to fast variables are described by the equations for oscillating elements. It was possible to make analogous corrections for the problems considered in Sec. 12.3, 12.4, and so on, for refinement of an automobile suspension model in Sec. 22.1.

Now a refined model of an aircraft angular motion can be constructed. Equations (21.10) are normalized for this class of motion and contain small parameters μ_2, ε_1, ε_2. The equations of the zeroth approximations with respect to these parameters are given by the system (21.11). Refine the approximate solution according to the Poincaré approach, choosing the largest small parameter μ_3 to be the small parameter of the expansion.

Two terms of the expansion should be enough:

$$v_{(1)} = v^{(0)} + \mu_3 v^{(1)}, \qquad \theta_{(1)} = \theta^{(0)} + \mu_3 \theta^{(1)}, \dots \tag{21.18}$$

The equations of zeroth approximation for (21.10) can be written in the form

$$\begin{aligned}
&m^{(0)} \frac{dv^{(0)}}{dt} = 0, \qquad m^{(0)} v^{(0)} \frac{d\theta^{(0)}}{dt} = 0,\\
&i_{zz}^{(0)} \frac{d\omega_z^{(0)}}{dt} = \left[m_z^{\alpha} \alpha^{(0)} + m_z^{\delta_z} \delta_z \right] \rho^{(0)} v^{(0)2},\\
&\frac{d\vartheta^{(0)}}{dt} = \omega_z^{(0)}, \qquad \vartheta^{(0)} = \theta^{(0)} + \alpha^{(0)},\\
&\frac{dm^{(0)}}{dt} = 0, \quad \frac{dy^{(0)}}{dt} = 0, \quad \frac{di_{zz}^{(0)}}{dt} = 0, \quad \rho^{(0)} = \rho(y^{(0)}).
\end{aligned} \tag{21.19}$$

The equations of the first approximation with respect to μ_3 are composed according to the general rule from Sec. II.

As before, the refined model (21.18) for the system (21.14) is described using the cumbersome two-step technique. Let us try to simplify it.

It should be taken into consideration that from Eqs. (21.19) it follows that $v^{(0)}$, $\theta^{(0)}$, $m^{(0)}$, $i_{zz}^{(0)}$, $\rho^{(0)}$ = const. If in the small components of the system (21.10) having order of μ_1, μ_2, ε_1, ε_2, we now substitute slow variables v, θ, m, i_{zz}, ρ by their zeroth approximations, then the system of equations obtained will differ from (21.10) by values of the second order. Let us write the system changed by this method, using new notations for variables $\widetilde{v}$, $\widetilde{\theta}, \dots$:

$$\begin{aligned}
m^{(0)} \frac{d\widetilde{v}}{dt} &= \mu_3 (-m^{(0)} \sin\theta^{(0)} + p - \rho^{(0)} v^{(0)2} c_x(\widetilde{\alpha})),\\
m^{(0)} v^{(0)} \frac{d\widetilde{\theta}}{dt} &= \mu_3 [-m^{(0)} \cos\theta^{(0)} + (p + \rho^{(0)} v^{(0)2} c_y^{\alpha}) \widetilde{\alpha}],\\
i_{zz}^{(0)} \frac{d\widetilde{\omega}_z}{dt} &= (m_z^{\alpha} \widetilde{\alpha} + m_z^{\delta_z} \delta_z) \rho \widetilde{v}^2\\
&\quad + \Big[(m_z^{\omega_z} + m_z^{\dot\alpha}) \varepsilon_1 \frac{\widetilde{\omega}_z}{v^{(0)}} - m_z^{\dot\alpha} \frac{\varepsilon_2}{m^{(0)} v^{(0)2}}\\
&\quad \times (-m^{(0)} \cos\theta^{(0)} + (p + \rho^{(0)} v^{(0)2} c_y^{\alpha}) \widetilde{\alpha}) \Big] \rho^{(0)} v^{(0)2},\\
\frac{d\widetilde{\vartheta}}{dt} &= \widetilde{\omega}_z, \quad \widetilde{\vartheta} = \widetilde{\theta} + \widetilde{\alpha}, \quad \lambda \frac{d\widetilde{y}}{dt} = \mu_1 v^{(0)} \sin\theta^{(0)}, \quad \rho = \rho(\widetilde{y}).
\end{aligned} \tag{21.20}$$

The system of Eqs. (21.20) is regarded as the refined model of angular aircraft motion.

By analogy with the system (21.14), the approximate solutions for the systems (21.10) and (21.20) coincide with accuracy up to the first order: $v_{(1)} = \widetilde{v}_{(1)}$, $\theta_{(1)} = \widetilde{\theta}_{(1)}, \dots$. As well as (21.17), the system (21.20) can be formed if the equations of the first approximation are multiplied by μ_3, added to the corresponding equations of the zeroth approximation (21.19), and proper transformations with an error of μ_3^2 are carried out.

Equations (21.20) are simpler than initial ones (21.10). The first equation in (21.20) is weakly connected with the other ones by the quadratic dependence of the frontal resistance c_x on the angle of attack $\widetilde{\alpha}$. Let us point out the important motion classes for which this equation is split.

Let a flight proceed with the small angles of attack $\widetilde{\alpha} \sim \mu_3$. Then with the error of order μ_3^2 in (21.20) it can be accepted that $c_x(\widetilde{\alpha}) = \text{const}$. Let us assume that for large values of $\widetilde{\alpha}$, corresponding to aerobatics, the variable $\widetilde{\alpha}$ can be expressed as a sum of a programmed value $\widetilde{\alpha}_{\text{n}}(t)$ and a deviation from it: $\widetilde{\alpha} = \widetilde{\alpha}_{\text{p}}(t) + \Delta\widetilde{\alpha}$. Let the deviations from a program be small $\Delta\widetilde{\alpha} \sim \mu_3$. Then with the error of order μ_3^2 it can be assumed that $c_x = c_x(\widetilde{\alpha}_{\text{p}}(t))$.

In both cases according to the first equation of (21.20), it can be assumed that the variable $\widetilde{v}$ is an explicit function of time with the error of order μ_3^2. In this case the second, third, and fourth equations form a closed system of second-order equations.

The value ρ in the third equation is defined by the two last equations in (21.20). According to the table in Sec. 21.1 for all aircraft types $\mu_1 \ll \mu_3$; therefore, $\rho = \text{const}$ is usually accepted. The influence of variability of ρ is easy to estimate by (21.20).

For the case of steady-state flight mode $\widetilde{v} = v^{(0)} = \text{const}$, $\rho = \text{const}$, $\theta^{(0)} = 0$ the equation structure of the refined model of an angular aircraft motion (21.20) is similar to the equation structure from Byushgens and Studnev [74] and Bochkarev [75] obtained by qualitative considerations.

22 Automobile motion decomposition

An automobile is a complicated dynamic system, comprised of a set of interacting subsystems: a body, a suspension, wheels, an engine, a transmission, controls. Any trial to construct "complete" motion equations is a notoriously dead-end task because of its excessive awkwardness.

The matter is simplified by the circumstance that in an automobile motion it is possible to separate those components significantly different with respect to the hierarchy of their time scales. Thus, the most clearly differentiated are:

(1) long-periodical, "route" motion components having characteristic times of the order of tens of minutes;

(2) mid-periodical trajectory motions in transient modes such as banking, acceleration, and deceleration, with characteristic times of the order of tens of seconds;

(3) short-periodical elastic oscillations of a body and a suspension with characteristic times of the order of second fractions [78, 79].

In the practice of automobile dynamics computation, simplified mathematical models have been established to describe a motion in terms of the specific time scale components. For justification of these simplified models only considerations of qualitative character or, at best, data of a comparison with an experiment or with the results of computations using more complicated models have been presented.

Thus, a problem is outlined — the problem of automobile motion separation, close to one of flight dynamics. We use the methods of fractional analysis to solve this problem.

22.1 Structure of automobile motion partial models

Any detailed description of an automobile dynamic is beyond the scope of frames of this book and the aims it pursues. Therefore, we limit the discussion to a description of the structure of the main equations of the system, to an estimation of the order of its partial time constants, and to the determination of a structure of its partial models of motion.

As a preliminary to the choice of notation for a coordinate system and variables of the problem it should be noted that in the automotive literature a uniformity for them has not been established. The aerodynamic perturbations with their stable notation system are also essential for an automobile, and we also use the notations of flight dynamics. Thereby the x axis is directed forward, the y axis up, and the z axis to the right with respect to the motion direction.

Let us introduce the right orthogonal trihedral $O_n x_n y_n z_n$, determined by the long-wave component of a road profile. Its origin O_n is placed on the middle line of an automobile motion band. The axis $O_n x_n$ is tangential to the middle line, and the axis $O_n y_n$ is normal to the road surface in the point O_n. The point O_n is positioned in such a way that the automobile mass center C lies in the plane $O_n y_n z_n$. The route velocity vector $\vec{V}$ of the point C is given in the axes $O_n x_n y_n z_n$ by its projections V_{x_n}, V_{y_n}, V_{z_n}. A distance L, covered by an automobile, is measured along the middle line and defined by the equation

$$\frac{dL}{dT} = V_{x_n}. \tag{22.1}$$

The length count of L begins from the initial position $O_o x_n y_n z_n$ of the trihedral $O_n x_n y_n z_n$.

An orientation of the trihedral $O_n x_n y_n z_n$ relative to $O_o x_n y_n z_n$ is determined by the turning angles Ψ_n, Θ_n, Γ_n relative to the consecutive positions of the axes y, z, x. Setting the relations $\Psi_n = \Psi_n(L)$, $\Theta_n = \Theta_n(L)$, $\Gamma_n = \Gamma_n(L)$ determines the long-wave component of the roadbed profile.

The short-wave profile components (microprofile) are determined by the functions $Y_n^{ij} = Y^{ij}(X_n^{ij}, Z_n^{ij})$. Here $i = 1, 2, \dots$ is the index of an automobile axis, to count from the front to the back, $j = 1, 2$ are indices of the left side and right side of an automobile relative to the motion direction, Y_n^{ij} is a height of a point of support for the wheel with indices i and j above the plane $O_n x_n z_n$, and X_n^{ij}, Z_n^{ij} are the coordinates of this point in the coordinate system used.

The equations of the automobile trajectory motion are written in the form of the equations for momentum change in the axes x_n, z_n of the trihedral $Cx_n y_n z_n$, obtained by the parallel transposition of the trihedral $O_n x_n y_n z_n$ into the automobile mass center

$$\frac{dQ_{x_n}}{dT} = \sum_{i,j} P_{x_n}^{ij} + G_{x_n}^a, \quad \frac{dQ_{z_n}}{dT} = \sum_{i,j} P_{z_n}^{ij} + G_{z_n}^a. \tag{22.2}$$

Here Q_{x_n}, Q_{z_n} are projections of the vector of the momentum of the automobile as a whole; $P_{x_n}^{ij}$, $P_{z_n}^{ij}$ are projections of the tangent forces of the interaction of wheels with a road which are external for the system considered; $G_{x_n}^a$, $G_{z_n}^a$ are projections of gravity forces, inertia forces, and aerodynamic forces applied to an automobile as a whole. Aerodynamic forces are written in the form accepted in the flight dynamics [73–75], and they are functions of the angles of slide β and attack α of the automobile body. The orientation of the body and trihedral $Cxyz$ connected to the body relative to $O_n x_n y_n z_n$ is determined by the angles of slide, attack, and roll ψ, ϑ, γ.

The translational and angular oscillations of the body caused by an elastic force of a suspension is written using an equation for motion of the body mass center in the axis y_n of the trihedral $Cx_n y_n z_n$ and equations for kinetic moment in the axes x, z of the trihedral $Cxyz$:

$$\begin{aligned} M\frac{d^2Y}{dT^2} &= \sum_{i,j}(C^{ij}\Delta Y^{ij} + R^{ij}\frac{d}{dT}\Delta Y^{ij}) + G_{y_n}^b, \\ I_x\frac{d^2\gamma}{dT^2} &= \sum_{i,j} -(C^{ij}\Delta Y^{ij} + R^{ij}\frac{d}{dT}\Delta Y^{ij})Z_c^{ij} + M_x^b, \\ I_z\frac{d^2\vartheta}{dT^2} &= \sum_{i,j}(C^{ij}\Delta Y^{ij} + R^{ij}\frac{d}{dT}\Delta Y^{ij})X_c^{ij} + M_z^b. \end{aligned} \tag{22.3}$$

Here M is the body mass; Y is the body mass center coordinate; C^{ij}, R^{ij} are reduced coefficients of stiffness and damping of a suspension; ΔY^{ij} are the values of the reduced spring deformations; $G^b_{y_n}$, M^b_x, M^b_z are projections of the principal vector and the principal moment of other forces applied to the body; I_x, I_z are the body moments of inertia with respect to the axes mentioned; X^{ij}_c, Z^{ij}_c are the coordinates of the reduced points of application of forces developed by the suspension. Equations (22.3) are written for simplicity using a number of assumptions: the trihedral $Cxyz$ coincides with the main axes of inertia of the body, the angles γ, ϑ are small, and so on. From the motion separation point of view these assumptions are not essential.

The equation for an angular automobile motion in the plane of the trajectory motion under the same simplifying assumptions is writtten in the form of the equation for the angular momentum with respect to the axis Cy_n of the trihedral $Cx_ny_nz_n$

$$I^a_y \frac{d^2\psi}{dT^2} = \sum_{i,j}(P^{ij}_x Z^{ij}_c - P^{ij}_z X^{ij}_c + L^{ij}_{y_n}) + M^a_{y_n}. \tag{22.4}$$

Here I^a_y is the reduced moment of inertia of the automobile as a whole; $M^a_{y_n}$ is the projection of the moment of all the forces except for contact ones; $L^{ij}_{y_n}$ is the moment of the contact friction of turning. In (22.4) it is assumed for simplicity that the coordinates in the axes x and z of the points of application of contact forces and suspension forces from (22.3) coincide.

The equations of the small vertical oscillations in the system "wheel+suspension" have the following structure

$$M^{ij}\frac{d^2Y^{ij}}{dT^2} = -C^{ij}\Delta Y^{ij} - R^{ij}\frac{d}{dT}\Delta Y^{ij} + N^{ij}_{y_n} + G^{ij}_{y_n}. \tag{22.5}$$

Here in addition to the previous notations, M^{ij} is the mass of the system under consideration reduced to the coordinate Y^{ij} which is the displacement along the axis O_ny_n of its mass center; $N^{ij}_{y_n}$ is the normal reaction in the point of the wheel contact with the supporting surface; $G^{ij}_{y_n}$ are projections of the other forces.

Suppose the deformed part of the pneumatic is to be localized in a small vicinity of a contact point. The motion of this part of the pneumatic for each of the wheels is described by the equations of momentum in the axes x_n, z_n and the angular momentum relative to the axis parallel to O_ny_n and passing through the contact point

$$\begin{aligned}
\frac{d\Delta Q^{ij}_{x_n}}{dT} &= F^{ij}_{x_n} + P^{ij}_{x_n}, \\
\frac{d\Delta Q^{ij}_{z_n}}{dT} &= F^{ij}_{z_n} + P^{ij}_{z_n}, \\
\frac{d\Delta K^{ij}_{y_n}}{dT} &= M^{ij}_{y_n} + L^{ij}_{y_n}.
\end{aligned} \tag{22.6}$$

Here $\Delta Q^{ij}_{x_n}$, $\Delta Q^{ij}_{z_n}$, $\Delta K^{ij}_{y_n}$ are components of the vector of momentum and vector of angular momentum for the system under consideration. By $F^{ij}_{x_n}$, $F^{ij}_{z_n}$, $M^{ij}_{y_n}$ components of the principal vector and principal moment of forces of interaction of the system with the not-deformed part of the pneumatic are designated.

Equations (22.1) through (22.6) do not form a closed system of equations. To close it the system should be supplemented with the necessary geometric and kinematic equations, equations describing the engine, controls, and the like. But it is possible to avoid this: to reveal a possibility of a motion separation the preceding equations are sufficient.

Let us normalize Eqs. (22.1) through (22.6):

$$\begin{gathered} t = \frac{T}{T_*}, \quad l = \frac{L}{L_*}, \quad v_{x_n} = \frac{V_{x_n}}{V_*}, \quad q_{x_n} = \frac{Q_{x_n}}{Q_*}, \dots \\ p^{ij}_{x_n} = \frac{P^{ij}_{x_n}}{P_*}, \quad f^{ij}_{x_n} = \frac{F^{ij}_{x_n}}{F_*}, \quad \dots, \quad \Delta q_{x_n} = \frac{\Delta Q_{x_n}}{\Delta Q_*}, \dots . \end{gathered} \tag{22.7}$$

After dividing each of the equations in (22.1) through (22.6) by a factor of corresponding equation dimension, the following partial time constants can be introduced:

$$\begin{gathered} T_0 = \frac{L_*}{V_*}, \qquad T_1 = \frac{Q_*}{P_*}, \qquad T_2 = \sqrt{\frac{M}{C_*}}, \\ T_3 = \sqrt{\frac{I_y \psi_*}{P_* X^{ij}_*}}, \quad T_4 = \sqrt{\frac{M^{ij}_*}{C_*}}, \quad T_5 = \frac{\Delta Q_*}{P_*}. \end{gathered} \tag{22.8}$$

In (22.8) it is assumed that $Q_* = MV_*$, $I_y = M\rho_y^2$, where ρ_y is the radius of inertia; $\Delta Q_* = \Delta M_* V_{0*}$, where ΔM_* is the characteristic mass of the deformed part of the pneumatic, and V_{0*} is the characteristic velocity of sliding upon the supporting surface. We assume all forces acting in the systems under consideration to be of one order $P_* = F_* = C_* \Delta Y_* = Mg$. It is natural to assume that $\Delta M_* \ll M^{ij}_* \ll M$. For estimating the partial time constants from (22.8) we suppose $L_* = 10 \div 10^2$ km, $V_* = 10 \div 10^2$ km/h, $M = 10^3$ kg, $M^{ij}_* = 50$ kg, and $\Delta M_* = 5$ kg. Assuming the possibility of the full sliding we accept $V_{0*} = V_*$. Let us suppose $\rho_y \sim X^{ij}_* = 3$ m. We choose the characteristic suspension stiffness C_*, providing the static deformation $\Delta Y_* = 0.1$ m. Then C_* is calculated from the relation $C_* \Delta Y_* = Mg$. The value ψ_* we set to be equal to the angle of the static turn of the unmoving automobile under the action of a moment of order $P_* X^{ij}_*$. Assume ψ_* to have the value of an order of some arc degrees.

Let us substitute parameter values in (22.7), chosen using these quite rough estimates. We obtain $T_0 \sim 10^3 \div 10^4$ sec, $T_1 \sim 0.3 \div 3$ sec, $T_2 \sim 10^{-1}$ sec, $T_3 \sim 10^{-1}$ sec, $T_4 \sim 0.3 \cdot 10^{-1}$ sec, and $T_5 \sim 10^{-2}$ sec.

Significant difference between numerical values of partial time constants makes it possible to form partial models of automobile motion. For this purpose, by analogy with Sec. 21, the value of T_* in (22.7) is set to be equal to the time constant corresponding to the motion class which is separated for approximate description. After the introduction of small parameters corresponding to this class, one of a variety of the small parameter techniques should be used.

Remark. Equations of a partial model can be obtained without carrying out a detailed procedure of fractional motion separation. To do this in the initial equations, all variables, which are slower than singling out, are assumed to be parameters, and the equations with respect to faster variables should be substituted by quasistatic relations.

22.2 Mathematical model of rolling for a deformed wheel. Are nonlinear nonholonomic constraints possible?

In the investigation of automobile dynamics the most ambiguous is, perhaps, a task of choosing a model describing an interaction of a wheel with a road. In the literature on automotive machines various models are discussed (Carter, Rocard, Keldysh, Fromm, ...), comparison of which is often impossible [78 – 82]. We consider a model of the interaction from which the models mentioned in the literature can be obtained as limit particular cases.

22.2.1 Let us suppose a wheel to consist of a rigid undeformable part (a disk) and deformable peripheral one (a pneumatic). A wheel is assumed to be not perturbed when its longitudinal symmetry plane is vertical, is parallel to the coordinate plane $C_n x_n y_n$, and the wheel is not loaded by external forces. The intersection of the outside pneumatic surface by its longitudinal symmetry plane is called "the middle line of the surface of rolling." If a tread design is disregarded, then in a unperturbed state the middle line will be a circle touching a support surface in the one point. For simplicity in the rest we assume the supporting surface to be a plane coinciding with the plane $O_n x_n z_n$. We designate by O the material point of a wheel coinciding in the current time moment with the point of contact. We set in the point O a linear element of the middle line connected to a pneumatic. The indices i, j for the vehicle wheels here and in the following are dropped.

Let us introduce the coordinate system $O_a x_a y_a z_a$. Its origin is placed into the point O_a of the intersection of a wheel axis with its symmetry plane, and the axis $O_a z_a$ is directed along the wheel axis. In the unperturbed state the trihedral $O_a x_a y_a z_a$ is oriented similarly to $C x_n y_n z_n$. The position of the point O with respect to the system $O_a x_a y_a z_a$ is set by the vector $\overline{R}$,

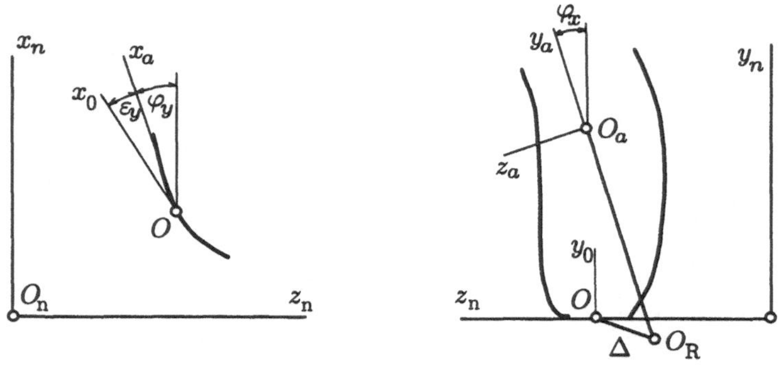

Fig. 48

oriented in the negative direction of the axis $O_a y_a$. Its length $R = \text{const}$ is a wheel radius.

Let us specify a perturbed disk position by two successive turns: one by the angle of φ_y with the axis $O_a y_a$ and another by the angle of φ_x with the intermediate position of the axis $O_a x_a$, by which the trihedral $O_a x_a y_a z_a$ is reset from an unperturbed state into a perturbed one (Fig. 48).

Let us load a wheel with external forces. In the perturbed state the point of contact becomes a spot. The perturbed position of the point O we call the center of the contact spot. We consider the wheel motion, in which the contact spot center remains on the supporting plane $O_n x_n z_n$. We introduce the trihedral $O x_o y_o z_o$ in the point O, with the axis $O x_o$, coinciding with the perturbed position of the middle line linear element in the point O, and with the axis $O y_o$, oriented along the normal to the supporting plane.

The perturbed position of the point O is given by the vector equation (Fig. 48)

$$\overline{O_a O} = \overline{O_a O_R} + \overline{O_R O}. \tag{22.9}$$

Here $\overline{O_a O_R} = \overline{R}$ is the vector previously introduced, connected to the disk in its unperturbed state together with the disk position, $\overline{O_R O} = \overline{\Delta}\,(\Delta_x, \Delta_y, \Delta_z)$ is the vector of the linear deformation of a pneumatic, and $\Delta_x, \Delta_y, \Delta_z$ are its projections on the axes of the trihedral $O x_o y_o z_o$.

The vector of a small turn $\overline{\varepsilon}\,(\varepsilon_x, \varepsilon_y, \varepsilon_z)$, bringing the trihedral $O_a x_a y_a z_a$ to the orientation similar to one of $O x_o y_o z_o$, determines the angular pneumatic deformation. Its projections on the axes x_o, y_o, z_o are designated by $\varepsilon_x, \varepsilon_y, \varepsilon_z$.

Let us define constraints applied for a motion of the trihedral $O x_o y_o z_o$. In wheel motion without detachment, but permitting a sliding relative to the supporting surface, the geometric conditions should be satisfied:

$$Y_{oy_n} = 0, \qquad \cos(y_o, x_n) = 0, \quad \cos(y_o, z_n) = 0 \tag{22.10}$$

and the kinematic conditions resulting from them

$$V_{oy_n} = 0, \qquad \Omega_{ox_o} = 0, \quad \Omega_{oz_o} = 0. \tag{22.11}$$

Here Y_{oy_n} and V_{oy_n} are the coordinate and velocity of the point O in the axes $Ox_ny_nz_n$; Ω_{ox_o} and Ω_{oz_o} are projections of the absolute angular velocity of the trihedral $Ox_oy_oz_o$ on its axes.

Let us write the equations for constraints (22.10) and (22.11) in explicit form. We eccept the transport assumptions of smallness of the values φ_x, φ_y, ε_y; $\dot\varphi_x$, $\dot\varphi_y$, $\dot\varepsilon_y$, traditional in the literature, and additional assumptions of smallness of ε_x, ε_z. The values $\dot\varepsilon_x$, $\dot\varepsilon_z$ could be not small as a result of significant angular deformations of a pneumatic caused by the running of a contact area around a wheel. The following calculations are done with accuracy of the first order with respect to the variables mentioned previously.

From the conditions in (22.10) for direction cosines it follows that

$$\varepsilon_x + \varphi_x = 0, \qquad \varepsilon_z = 0. \tag{22.12}$$

The vector $\overline{\omega}_o$ of the trihedral angular velocity $Ox_oy_oz_o$ is determined by the equation

$$\overline{\Omega}_o = \overline{\Omega}_a + \overline{\Omega} + \overline{\dot\varepsilon}, \tag{22.13}$$

where $\overline{\Omega}_a$ is the angular velocity vector of the trihedral $O_ax_ay_az_a$, $\overline{\Omega}$ is the angular velocity of the disk, and $\overline{\dot\varepsilon}$ is the angular velocity of deformation. In projections on the axes $x_oy_oz_o$ (22.13) can be written in the form:

$$\begin{aligned} \Omega_{ox_o} &= \dot\varphi_x - \Omega_{z_a}\varepsilon_y + \dot\varepsilon_x, \\ \Omega_{oy_o} &= \dot\varphi_y + \Omega_{z_a}\varepsilon_x + \dot\varepsilon_y, \\ \Omega_{oz_o} &= \Omega_{z_a} + \dot\varepsilon_z. \end{aligned} \tag{22.14}$$

Here Ω_{z_a} is the projection of the vector of the angular velocity of the disk on the axis z_a.

From (22.14) by taking into account (22.12) the following expressions for components of the angular velocity of deformation are obtained

$$\dot\varepsilon_x = -\dot\varphi_x + \Omega_{z_a}\varepsilon_y, \qquad \dot\varepsilon_z = -\Omega_{z_a}.$$

These values determine the energy dissipation in the material of a pneumatic caused by a deformation.

The second equation from (22.14) together with (22.12) determines the angular velocity of the trihedral $Ox_oy_oz_o$ turning

$$\Omega_{oy} = \dot\varphi_y + \dot\varepsilon_y - \Omega_{z_a}\varphi_x. \tag{22.15}$$

22.2.2 Now we get down to detailing the dynamic processes caused by a deformation, which were described previously by Eqs. (22.6) in the most general form.

Let us consider a fixed in its form, small in its thickness, geometric element adjacent to the supporting surface, rigidly connected with the trihedral $Ox_oy_oz_o$, and including completely the contact spot despite the possible change of its dimensions.

In the framework of Euler's formalism let us consider a mechanical system formed by material particles of a pneumatic, which in the given time moment are inside this geometric volume. Dynamic equations for this system, named the "supporting element," should be written.

An important remark should be made. The dynamic equations of the supporting element are used to describe route, trajectory, and oscillation classes of automobile motion. According to estimates from Sec. 22.1, the partial time constants $T_0, \ldots, T_4$ of these motion components are considerably greater than the time constant T_5, characterizing dynamic processes taking place in pneumatic deformation. Therefore, in the following the dynamic equations for the supporting element can be written in quasistatic form with the errors corresponding to the motion class $T_5/T_0, \ldots, T_5/T_4$.

In the projections on the axes of the trihedral $Ox_oy_oz_o$ we obtain

$$\begin{gathered} F_{x_o} + P_{x_o} = 0, \quad F_{y_o} + N = 0, \quad F_{z_o} + P_{z_o} = 0, \\ M_{x_o} - Z_N N = 0, \\ M_{y_o} + Z_N P_{x_o} - X_N P_{z_o} + L_{y_o} = 0, \\ M_{z_o} + X_N N + M_{kz} = 0. \end{gathered} \tag{22.16}$$

Here F_{x_o}, F_{y_o}, F_{z_o} and M_{x_o}, M_{y_o}, M_{z_o} are the components of the principal vector and principal moment of the elastic forces of interaction of the supporting element with the pneumatic body. As a point to reduce the forces, the point O is taken. As N is used to designate a value of the result of normal reaction forces from the the supporting surface, P_{x_o}, P_{z_o}, $L_{y_o}(O_N)$ are components of the principal vector and the principal moment of the system of tangent reaction forces; X_N, Z_N are the coordinates of the point O_N, which is the reduction point for forces of interaction with a supporting surface; and M_{kz} is a component of a rolling-friction moment. For M_{kz} we take the usual representation

$$M_{kz} = -\nu N \operatorname{sgn} \Omega_{z_a}, \tag{22.17}$$

where ν is a rolling-friction coefficient.

For a description of phenomena in the contact area ten additional variables are introduced: six deformation components: Δ_x, Δ_y, Δ_z; ε_x, ε_y, ε_z, two coordinates X_N, Z_N, and two reactions N and M_{kz}. They are connected by six equations (22.16), three constraint equations (22.11) and Eq. (22.17). Hence, this problem is a closed one. Of course in this case one should assume that the dependencies of forces and moments in (22.16) on geometric and kinematic variables are known.

22.2.3 We accept the assumption traditional for transport literature that a deformed state of a pneumatic is determined by a finite number of variables. As these quantities, we accept the components of the vectors of the linear $\overline{\Delta}$ and angular $\overline{\varepsilon}$ deformation from Sec. 22.2.1. Hence, the forces and moments $F_{x_o}, \ldots, M_{z_o}$ of the interaction between the supporting element and the pneumatic body depend on $\Delta_x, \ldots, \varepsilon_z$ and their derivatives $\dot{\Delta}_x, \ldots, \dot{\varepsilon}_z$.

For simplicity some additional assumptions should be accepted. We assume that forces and moments depend only on deformations, this relation is linear, and the matrix of the stiffness coefficients is diagonal. Then

$$\begin{aligned} F_{x_o} &= -F_x^x \Delta_x, \quad F_{y_o} = -F_y^y \Delta_y, \quad F_{z_o} = -F_z^z \Delta_z, \\ M_{x_o} &= -M_x^x \varepsilon_x, \quad M_{y_o} = -M_y^y \varepsilon_y, \quad M_{z_o} = -M_z^z \varepsilon_z, \end{aligned} \tag{22.18}$$

where $F_x^x, \ldots, M_z^z$ are the corresponding linearization coefficients.

Experimental data show that the values of forces and moments of interaction with the supporting surface P_{x_o}, P_{z_o}, L_{y_o} depend on arguments $s_x = V_{ox}/V_a$, $s_z = V_{oz}/V_a$, where V_o, V_a are the sliding velocity of the wheel and the velocity of its axis. These characteristics are continuous, limited, and pass through zero when s_x, s_z, $\Omega_{y_o} = 0$ [78, 79, 81]. For specific cases of motion analogous characteristics were obtained by Fromm [81] in the framework of the "brush-model." We estimate the characteristics in a more general case.

Let us consider a supporting element model, combining both hypotheses: of finite dimensionality of a deformed state of a pneumatic and Fromm's interaction with a supporting surface. A similar model of interaction of a rolling rigid body with a deformable supporting surface has been studied by Ishlinsky [82].

Let us assume that the supporting element consists of an undeformable part (a substratum) and an infinitely thin deformable boundary layer at the boundary with the supporting surface. We connect the previously introduced trihedral $Ox_oy_oz_o$ with the substrate. As the thickness of the boundary layer is supposed to be infinitely small, then the constraints (22.10) and (22.11), imposed on the trihedral $Ox_oy_oz_o$ motion remain valid. We suppose the boundary layer to be comprised of infinitely small noninteracting contact elements. Each of the elements is connected to the substrate elastically and interacts with the supporting surface according to Coulomb. A scheme for such a model of a supporting element for the case of one-dimensional motion along the axis x_o is introduced in Fig. 49.

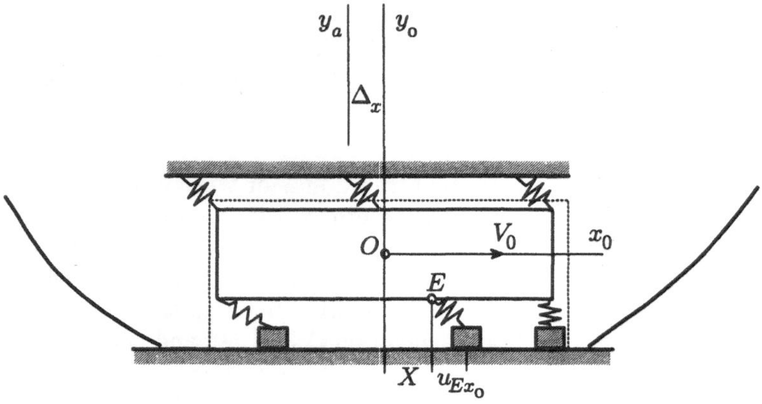

Fig. 49

Let us consider a single contact element and, within the framework of Lagrangian description, trace an evolution of forces acting on it. The supporting element distinguished in Fig. 49 by the dotted line is the system of varying composition. As a wheel rolls, the elements of the pneumatic periphery come into contact with the support surface, enter the supporting element volume, and then leave it.

Suppose that in the moment of the first touching of the surface by the contact element its velocity instantly becomes equal to zero, and its "spring" is undeformed. The point of attachment of the contact element spring to the substratum is designated by E. As the point E moves with the substratum the spring stretches. In the initial state, when the contact element has not begun to slip, an elastic force is acting on it from the substratum, which is balanced by a Coulomb friction of the support surface. We designate by $\overline{p}_E$ the vector of Coulomb force. Suppose the contact spot to be isotropic in its elastic properties. Then

$$\overline{p}_E = -k\overline{u}_E. \tag{22.19}$$

Here k designates the coefficient of stiffness per unit area of a contact spot, and $\overline{u}_E$ is two-component vector of a spring deformation.

A contact element begins to slip when the force $\overline{p}_E$ attains the limit value

$$|\overline{p}_E| = \mu_0 n. \tag{22.20}$$

Here μ_0 is the coefficient of the Coulomb friction of rest; n is a normal reaction per an area unit. After sliding begins, a friction force acts on the contact element

$$\overline{p}_E = -\mu_1 n \frac{\overline{V}_E}{V_E}, \tag{22.21}$$

where $\overline{V}_E$ is the velocity vector of a contact element, and $\mu_1 \leq \mu_0$ is the sliding friction coefficient.

The principal vector and the principal moment of forces applied to the set of contact elements by the support surface are given by the expressions

$$\overline{P} = \iint\limits_D \overline{p}_E \, dX_E dZ_E,$$
$$L_{y_o} = \iint\limits_D [\,\overline{p}_{EZ} X_E - \overline{p}_{EX} Z_E]\, dX_E dZ_E. \tag{22.22}$$

Here X_E, Z_E are the coordinates of the point E in the coordinate system $O_N x_o y_o z_o$ originating in the point O_N and with the axes in parallel with those of the trihedral $Ox_o y_o z_o$, and D designates the contact area.

Let us accept some simplifying assumptions.

We suppose the contact element mass and stiffness k to have such values that the characteristic of the time constant of its own motion is considerably less than all partial time constants of the system $T_0, \ldots, T_5$, which were estimated in Sec. 22.1. Then according to the usual reasoning of fractional analysis the motion equations of the contact element are degenerated into quasistatic relations, where the variables of all the hierarchically slower levels are assumed to be parameters.

The deformations u_E are supposed to be considerably less than the characteristic dimensions of a contact spot. Then while calculating (22.21) and integrating (22.22) the attachment points of the contact element spring can be indistinguishable.

Let us write the expressions (22.19) through (22.22) in the explicit form.

The spring deformation velocity of the contact element is determined by the difference of the velocities of its attachment points. According to Fig.49, the velocity of the "bottom" point is equal to zero, at least for some time after the first contact of this element with the surface. The velocity of the "top" point is equal to the absolute velocity of the point E. Therefore, in projections on the axes of the trihedral $Ox_o y_o z_o$ we have

$$\frac{du_{Ex_o}}{dT} = V_{Ex_o}, \qquad \frac{du_{Ez_o}}{dT} = V_{Ez_o},$$
$$u_{Ex_o}(0) = 0, \quad u_{Ez_o}(0) = 0, \tag{22.23}$$

where time is counted from the moment of the first touch. By Euler's theorem,

$$V_{Ex_o} = V_{Ox_o} + \Omega_{y_o} Z, \qquad V_{Ez_o} = V_{Oz_o} - \Omega_{y_o} X. \tag{22.24}$$

Here X, Z are the coordinates of the point E in the coordinate system $Ox_o y_o z_o$, which are connected with the coordinates X_E, Z_E of this point from (22.22) by the translation transformation

$$X = X_N + X_E, \qquad Z = Z_N + Z_E. \tag{22.25}$$

The expressions for projections V_{Ox_o}, V_{Oz_o} of the absolute velocity vector of the point O are obtained as a result of differentiation of Eq. (22.9) and proper calculations with the same as in Sec. 22.1, accuracy up to the terms of first order

$$\begin{aligned} V_{Ox_o} &= V_{ax_n} + \Omega_{z_a} R + \dot{\Delta}_x, \\ V_{Oz_o} &= V_{ax_n}(\varphi_y + \varepsilon_y) + V_{az_n} + \Omega_{z_a} R\varepsilon_y + \dot{\varphi}_x R + \dot{\Delta}_z. \end{aligned} \tag{22.26}$$

The changes in time of the values X, Z from (22.24) are determined by displacement of the point E relative to the coordinate system $O^e x_o^e y_o^e z_o^e$, coinciding in the successive time moments with the instantaneous positions of the trihedral $Ox_oy_oz_o$

$$\frac{dX}{dT} = V^r_{Ex_o}, \quad \frac{dZ}{dT} = V^r_{Ez_o}; \quad X(0) = X_D, \quad Z(0) = Z_D. \tag{22.27}$$

The initial conditions in (22.27) are connected with the equation of the boundary of the area D:

$$F(X_D, Z_D) = 0. \tag{22.28}$$

The projections $V^r_{Ex_o}$, $V^r_{Ez_o}$ reference-frame velocity of the point E relative to the coordinate system $O^e x_o^e y_o^e z_o^e$ is obtained from Eqs. (22.26), eliminating in them the reference-frame velocity components of the motion $O^e x_o^e y_o^e z_o^e$ together with the trihedral $O_a x_a y_a z_a$ caused by a deformation. Then

$$V^r_{Ex_o} = \Omega_{z_a} R, \qquad V^r_{Ez_o} = \Omega_{z_a} R\varepsilon_y. \tag{22.29}$$

In calculations (22.19) through (22.29), in accordance with the preceding remark, the slow variables V_{Ox_o}, V_{Oz_o}, Ω_{y_o}, Ω_{z_a}, ε_y, X_N, Z_N are supposed to be parameters.

Let us substitute (22.29) in (22.27) and integrate these equations. We obtain

$$X = \Omega_{z_a} RT + X_D, \qquad Z = \Omega_{z_a} R\varepsilon_y T + Z_D. \tag{22.30}$$

From (22.23), (22.24), and (22.30), it follows that

$$\begin{aligned} u_{Ex_o} &= (V_{Ox_o} + \Omega_{y_o} Z_D)T + \Omega_{y_o}\Omega_{z_a} R\varepsilon_y T^2/2, \\ u_{Ez_o} &= (V_{Oz_o} - \Omega_{y_o} X_D)T - \Omega_{y_o}\Omega_{z_a} RT^2/2. \end{aligned} \tag{22.31}$$

We eliminate in (22.31) the variable T using the first equation of (22.30)

$$\begin{aligned} u_{Ex_o} &= (V_{Ox_o} + \Omega_{y_o} Z_D)\frac{X - X_D}{\Omega_{z_a} R} + \Omega_{y_o}\varepsilon_y \frac{(X - X_D)^2}{2\Omega_{z_a} R}, \\ u_{Ez_o} &= (V_{Oz_o} - \Omega_{y_o} X_D)\frac{X - X_D}{\Omega_{z_a} R} - \Omega_{y_o} \frac{(X - X_D)^2}{2\Omega_{z_a} R}. \end{aligned} \tag{22.32}$$

Suppose that in (22.32) the wheel rotation agrees with the route velocity direction: $\Omega_{z_a} = -\Omega < 0$. We introduce notation:

$$s_x = V_{Ox_o}/\Omega R, \quad s_z = V_{Oz_o}/\Omega R, \quad \omega_{y_o} = \Omega_{y_o}/\Omega. \tag{22.33}$$

Then (22.32) will take the form:

$$\begin{aligned} u_{Ex_o} &= -(s_x + \omega_{y_o}\frac{Z_D}{R})(X - X_D) - \omega_{y_o}\varepsilon_y\frac{(X - X_D)^2}{2R}, \\ u_{Ez_o} &= -(s_x - \omega_{y_o}\frac{X_D}{R})(X - X_D) + \omega_{y_o}\frac{(X - X_D)^2}{2R}. \end{aligned} \tag{22.34}$$

The set of relations (22.19) through (22.34) permits the calculation of the areas of sliding and unsliding of the contact elements, the principal vector, and the principal moment of the forces acting from the rolling surface upon a supporting element.

Let us consider the two extreme cases of motion.

1. Suppose all contact elements do not slip. We substitute Eqs. (22.34) into (22.19) and then into (22.22). In integration in (22.22) the boundary of the area D and the values s_x, s_z, ω_{y_o} are supposed to be constant. It is seen that the values P_{x_o}, P_{z_o}, L_{y_o} are linear functions of arguments s_x, s_z, ω_{y_o}. The coefficients of these relations are determined by the form of the area D, the spot of contact.

These calculations could be completed, having accepted additional simplification.

Let us drop into (22.29) and (22.34) the components with the small factor ε_y. Suppose the area D to be a rectangle which is symmetrical relative to the axes of the trihedral $O_N x_o y_o z_o$, and has the lengths of the sides equal to $2a$ and $2b$. Then the successive positions of the point E in the supporting element are in the line parallel to the axis x_o, and in (22.34) it should be accepted that $X_D = X_N + a$, $Z_D = Z_N + Z_E$, $X - X_D = X_E - a$ according to (22.25). Calculations of (22.19), (22.34), and (22.22) produce

$$\begin{aligned} P_{x_o} &= -Ka(s_x + \frac{Z_N}{R}\omega_{y_o}), \\ P_{z_o} &= -Ka(s_z - \frac{X_N}{R}\omega_{y_o}) + K\frac{a^2}{3R}\omega_{y_o}, \\ L_{y_o} &= -K\frac{a^2}{3}(s_z - \frac{X_N}{R}\omega_{y_o}) - K\frac{ab^2}{3R}\omega_{y_o}. \end{aligned} \tag{22.35}$$

Here $K = 4abk$ is the total shear stiffness of a boundary layer, variables s_x denote the so-called longitudinal pseudosliding, s_z is the angle of lateral runoff and normalized angle velocity of rotation ω_{y_o}. These quantities are defined by (22.33).

2. Let us consider now a motion in which $\Omega_{y_o} \equiv 0$ and all contact elements slide. Then according to (22.24) the substratum is moving translationally and the system of Coulomb forces (22.21) is a parallel one. In addition let us suppose that the area D is symmetrical with respect to the axes $O_N x_o$, $O_N z_o$ and the distribution of normal reactions per area unit is also symmetrical. Then according to (22.21) and (22.22) we have $\overline{P} = -\mu_1 N \overline{V}_O / V_O$, $L_{y_o} = 0$, where N is the value of the principal vector of the forces of normal reactions. In the projection on the trihedral axes $Ox_o y_o z_o$,

$$P_{x_o} = -\mu_1 N \frac{s_x}{\sqrt{s_x^2 + s_z^2}}, \quad P_{z_o} = -\mu_1 N \frac{s_z}{\sqrt{s_x^2 + s_z^2}}, \quad L_{y_o} = 0. \tag{22.36}$$

From a comparison of (22.35) with (22.36) it follows that the characteristics $P_{x_o} = P_{x_o}(s_x)$, $P_{z_o} = P_{z_o}(s_z)$, $L_{y_o} = L_{y_o}(s_z)$ have linear parts (22.35) with respect to their arguments and they run into saturations (22.36), which fairly well agrees quantitatively with the data from Khachaturov [72], Ellis [79], and Pacejka [81].

22.2.4 Let us substitute into the quasistatic equations (22.16) the expressions (22.18) and (22.35) and the corresponding forces and moments for the linear parts of their characteristics. We get equations of the form:

$$\begin{gathered} -F_x^x \Delta_x - K a s_x = 0, \\ -F_z^z \Delta_z - K a s_z + K \frac{a^2}{3R} \omega_{y_o} = 0, \\ \dots . \end{gathered} \tag{22.37}$$

Let us obtain some qualitative conclusions depending on relations between the coefficients of Eqs. (22.37).

Let us use estimates of the form

$$\begin{aligned} F_x^x a(\Delta_{x*}/a) &= K a s_{x*} = \dots = G_*, \\ M_y^y \varepsilon_{y*} &= L_y^\omega \omega_{y_o *} = M_*^a, \end{aligned} \tag{22.38}$$

where Δ_{x*}, $s_{x*}, \dots, \omega_{y_o *}$ are the characterizing values of the corresponding variables, and G_*, M_*^a are the characterizing values of external forces and moments applied to an automobile.

Consider the case of "large" coefficients $F_x^x a$, $Ka, \dots, M_y^y$, so that for limited values of G_*, M_*^a from (22.38) it follows that Δ_{x*}/a, $s_{x*}, \dots, \varepsilon_{y*} \ll 1$. For $F_x^x a$, $Ka, \dots, M_y^y \to \infty$ we have Δ_{x*}/a, $s_{x*}, \dots, \varepsilon_{y*} \to 0$. Naturally, the limitary relations for characteristic values gained result in analogous expressions for the variables themselves: Δ_x/a, $s_x, \dots, \varepsilon_y \to 0$.

The limitary relations for Δ_x, Δ_z, $\varepsilon_y \to 0$ give the condition of undeformability of a wheel, and the relations for s_x, $s_z \to 0$ for the wheel with a finite velocity $R\Omega \neq 0$ give, according to (22.33), the conditions for not sliding.

If the set of limitary conditions for the coefficients of Eqs. (22.37) is supplemented with the condition $L_y^\omega \to \infty$, then from (22.38) follows the limitary relation $\omega_{y_o *} \to 0$. In this case there is no turning of the wheel about the vertical axis.

An automobile having four or more supporting wheels is a kinematically indefinable system according to Sec. 17. Conditions of not sliding for all supporting wheels give an overdefined set of equations. In this case the set of quasistatic equations similar to (22.37) for all wheels should be included in the set of equations for the automobile model, as was done in Sec. 17.3.

Having substituted in (22.37) the expressions (22.33) and (22.26) and assuming for now the coefficients of stiffness $F_x^x a, \ldots$ and friction Ka to be the values of equal order, we obtain

$$\begin{aligned}
&-F_x^x a(\Delta_x/a) - Ka(V_{ax_n} - \Omega R + \dot{\Delta}_x)/\Omega R = 0,\\
&-F_z^z a(\Delta_z/a) - Ka(V_{ax_n}(\varphi_y + \varepsilon_y) + V_{az_n}\\
&\qquad -\Omega R\varepsilon_y - \dot{\varphi}_x R + \dot{\Delta}_z) + K\frac{a^2}{3R}\omega_{y_o} = 0,\\
&\ldots .
\end{aligned} \tag{22.39}$$

The values V_{ax_n}, $\Omega, \ldots$ are expressed in Sec. 22.1 in terms of variables of the upper hierarchical automobile levels.

The equations of the first order from (22.39) connect the automobile variables and their derivatives. Generally speaking, these equations are nonintegrable; that is, they cannot be reduced to explicit relations between the variables. For the specific case of a motion considered by Keldysh [48, 80], the so-called equations of holonomic constraints of the Keldysh theory of rolling have been obtained from (22.38).

We now consider the case of $F_x^x a \gg Ka, \ldots$. Then passing to a limit with respect to coefficients of stiffness $F_x^x a \ldots \to \infty$ we obtain, as in Sec. 16, a model of an absolutely rigid wheel $\Delta_x, \ldots \to \infty$. Then from (22.16) and (22.39) with the additional limitation $\omega_{y_o} = 0$ the expressions for the forces of interaction of a wheel with a road are obtained:

$$\begin{aligned}
F_{x_o} &= -Ka(V_{ax_n} - \Omega R)/\Omega R,\\
F_{z_o} &= -Ka(V_{az_n} + V_{ax_n}\varphi_y - \dot{\varphi}_x R)/\Omega R,
\end{aligned}$$

corresponding to the a priori models of Roccard and Carter [80].

Let us accept unchanged all the limitations of the previous cases, except one. We assume the magnitude of the force P calculated in accordance with (22.22) has reached its maximal value. After this — by analogy with

Sec. 15.2, 15.3, and 17.2 — we come to the model of an absolutely rigid wheel, rolling with sliding with finite values of the s_x and s_z.

Remark. Equations (22.39) are written for the linear parts of force characteristics from (22.16). After a transition to nonlinear parts of the characteristics, the quasistatic equations of the form of (22.16) become nonlinear. With the same etymological argumentation as (22.39), they can be named the equations of unholonomic but, in this case, nonlinear constraints.

22.2.5 In classical mechanics the opinion has been established, almost on the level of a postulate, about the impossibility of nonlinear unholonomic constraints. Thus the statement just delivered should be discussed especially.

In classical mechanics unholonomic constraints are imposed on a system with given dimensionality of configuration space. The imposing of these constraints does not change the space dimensionality and does not change the matrix of inertia coefficients determining the space metric.

In classical mechanics the linearity of holonomic constraints is determined by the circumstance that the constraints discussed in the literature have a kinematic character, that is, prohibitions on some kind of mutual motions of bodies. Due to linearity with respect to velocities of the theorem about velocity composition any restrictions on relative velocities have linear structure.

In the problem of a deformed-wheel rolling the set of generalized coordinates for a supporting element increases the dimensionality of the configurational space of the system with a rigid wheel. The singular neglect of the supporting element mass degenerates the matrix of the inertia coefficients of the system. Thus, quasistatic equations obtained in the form (22.16) can be treated as equations of constraints imposed on the system in the expanded configuration space with metrics defined by the confluent matrix. The linearity or nonlinearity of these quasistatic equations is absolutely not the circumstance of principle; it depends only on a form of force characteristics.

The similar situations in the electro- and radiophysics were known long ago [8]. There the degeneration with respect to small inductances results in the models with "half" degrees of freedom, which are the direct analogues of the preceding models.

In mechanics the typical example of relations of the unholonomic kind, obtained as a result of degeneration of a matrix of coefficients of inertia, can be precession relations. Notions such as "unholonomic constraint" and "unholonomic motion" in precession formulations are widely used in the literature on gyro theory [15].

References

1. Kline S.Z., *Similitude and Approximation Theory*, McGraw-Hill Book Company, New York, 1965.

2. Sedov L.I., *Dimensional and Similarity Methods in Mechanics* (trans.), Academic Press Inc., New York, 1960.

3. Novozhilov I.V. *Approximate Methods of Dynamical Systems Investigation*, Moskovsk. energet. inst., Moscow, 1980 (in Russian).

4. Novozhilov I.V., *Fractional Analysis*, Moskovsk. gos. univ., Moscow, 1991; Moscow, 1995 (in Russian).

5. Beletsky V.V., *Essays About Motion of Celestial Bodies*, Nauka, Moscow, 1977 (in Russian).

6. Novozhilov I.V., Approximate Methods of Research of Gyro-Systems, in: *Development of Mechanics of Gyro and Inertial Systems*, Nauka, Moscow, 1973, pp. 368–377 (in Russian).

7. Vasil'eva A.B. and Butusov V.F., *Asymptotic Methods in Singular Perturbation Theory*, Vysshaya shkola, Moscow, 1990 (in Russian).

8. Andronov A.A., Vitt A.A., and Khaykin S.E., *Theory of Oscillations*, Fizmatgiz, Moscow, 1959 (in Russian).

9. Wasow W., *Asymptotic Expansions for Ordinary Differential Equations*, Interscience Publishers, a division of John Wiley & Sons, New York. London. Sydney, 1965.

10. Blekhman I.I., Myshkis A.D., and Panovko Ya.G., *Applied Mathematics: Matter, Logic, Features of Approaches*, Naukova dumka, Kiev, 1976 (in Russian).

11. Zharkov V.N., *Inner Texture of Earth and Planets*, Nauka, Moscow, 1983 (in Russian).

12. Mischenko E.F. and Rozov N.Kh., *Differential Equations with Small Parameter and Relaxation Oscillations*, Nauka, Moscow, 1975 (in Russian).

13. Moiseev N.N., *Asymptotic Methods of Nonlinear Mechanics*, Nauka, Moscow, 1981 (in Russian).

14. Kuz'mina R.P., *Method of Small Parameter in Regularly Perturbed Cauchy Problem* Moskovsk. Gos. Univ., Moscow, 1991 (in Russian).

15. Ishlinsky A.Yu., *Mechanics of Gyroscopic Systems*, Akad. Nauk SSSR, Moscow, 1963 (in Russian).

16. Novozhilov I.V., Dependence of Drifts of Three-Axis Gyroscopic Platform on Disposition of Gyroscopes on the Platform, *Inzhen. zhurn. Mekhan. tverd. tela*, No. 2, 1968, pp. 26–29.

17. Novozhilov I.V., Dependence of Drift of Gyroscopic Devices on Zeroth Root Degree of Linear Part of Equations, *Izv. Akad. Nauk SSSR, Mekhan. tverd. tela*, No. 4, 1971, pp. 38–42

18. Malkin I.G., *Some Problems of Nonlinear Oscillations Theory*, GITTL, Moscow, 1956 (in Russian).

19. Bogolyubov N.N. and Mitropolsky Yu. A., *Asymptotic Methods in Theory of Nonlinear Oscillations*, GIFML, Moscow, 1958 (in Russian).

20. Grebennikov E.A., Averaging Methods in Applied Problems, Nauka, Moscow, 1986 (in Russian).

21. Volosov V.M. and Morgunov B.I., *Averaging Method in the Theory of Nonlinear Oscillating Systems*, Moskovsk. Gos. Univ., Moscow, 1971 (in Russian).

22. Zhuravlev V.F. and Klimov D.M., *Applied Methods in the Oscillations Theory*, Nauka, Moscow, 1988 (in Russian).

23. Popov E.P. and Pal'tov I.P. *Approximate Methods of Investigation of Nonlenear Control Systems*, Fizmatgiz, Moscow, 1980 (in Russian).

24. Bulgakov B.V., *Oscillations*, GITTL, Moscow, 1954 (in Russian).

25. Tikhonov A.N., Systems of Differential Equations Containing Small Parameters by Derivatives, *Matem. sbornic*, Vol. 31(73), No. 3, 1952, pp. 575–586.

26. Vasil'eva A.B. and Butusov V.F., *Asymptotic Expansions of Singular Perturbed Systems Solutions*, Nauka, Moscow, 1973 (in Russian).

27. Gradstein I.S., About Solutions of Differential Equations with Small Parameters by Derivatives on Time Half-Line, *Matem. sbornik*, Vol. 32, 1953, pp. 533–544.

28. Krasovsky N.N. and Klimushev A.I., Uniform Asymptotic Stability of the Differential Equations Systems with Small Parameter by Higher Derivatives, *Uspekhi matem. nauk*, Vol. 18, no. 3, 1963, pp. 680–690.

29. Lomov S.A., *Introduction to the General Theory of Singular Perturbations*, Nauka, Moscow, 1981 (in Russian).

30. Kus'mina R.P., *Method of Small Parameter for Singularly Perturbed Systems*, Moskovsk.gos. univ., Moscow, p.1 1993, p.2 1994 (in Russian).

31. Bagryantsev V.I., Nevarko V.S., and Rabinovich B.I., Mathematical Model of Car with Electromagnetic Suspension, Izv. Akad. Nauk SSSR, Energetika i transport, No. 1, 1981, pp. 101–107.

32. Novozhilov I.V., Motions Decomposition in the Problem About the Train with Magnetic Suspension, *Trudy Moskovsk. energet. inst.*, No. 573, 1981, pp. 3–8.

33. Kapustina O.M. and Novozhilov I.V., Motions Decomposition in Dynamics of Vehicle with Contactless Suspension, *Mezhvuzovsk. temat. sbornik* No. 16, Moskovsk. energet. inst., 1983, pp. 36–42.

34. Bragin V.V., Novozhilov I.V., and Pshenichkina L.A., About Stability of Three-Axis Indicating Gyroscopic Stabilizer, *Izv. Akad. Nauk SSSR, Mekhan. tverd. tela*, No. 6, 1969, pp. 26–33.

35. Zatsepin M.F. and Novozhilov I.V., Control of Quadruped Gaits, *Izv. Akad. Nauk SSSR, Mekhan. tverd. tela*, No. 5, 1986, pp. 60–66.

36. Filippov A.F., *Differential Equations with Discontinuous Right-hand Sides*, Kluwer Academic Publishers, Boston, 1988.

37. Utkin V.I., *Sliding Modes and Their Applications in the Systems with Variable Structure*, Nauka, Moscow, 1974 (in Russian).

38. Gerashchenko E.I. and Gerashchenko C.M., *Method of Motions Decomposition and Optimisation of Nonlinear Systems*, Nauka, Moscow, 1975 (in Russian).

39. Novozhilov I.V., Stagnation Conditions in Systems with Coulomb Friction, *Izv. Akad. Nauk SSSR, Mekhan. tverd. tela*, No. 1, 1975.

40. Boltyansky V.G. and Pontryagin L.S., About Equilibrium State Stability of "Relay" System of Ordinary Differential Equations, *Trudy 3-go Vsesoyuzn. matem. s'ezda, Izd. AN SSSR*, Vol. 1, 1956, pp. 217–218.

41. Tsypkin Ya.Z., *Theory of Relay Automatic Control Systems*, Gostekhizdat, Moscow, 1955 (in Russian).

42. Alekseev K.B. and Bebenin G.G., *Control of Spacecraft*, Mashinostroenie, Moscow, 1974 (in Russian).

43. Novozhilov I.V., Precession Equations of Gyroscopic Systems with "Stiff" Control with Respect to Some Variables, *Izv. Akad. Nauk SSSR, Mekhan. tverd. tela*, No. 1, 1971, pp. 144–148.

44. Novozhilov I.V., About Correctness of Some Limit Models in Mechanics, *Preprint of Inst. probl. mekhan. AN SSSR*, No. 253, 1985.

45. Chernous'ko F.L., Dynamics of System with Elastic Elements of High Stiffness, *Izv. Akad. Nauk SSSR, Mekhan. tverd. tela*, No. 4, 1983, pp. 101–113.

46. Novozhilov I.V., Limit Model of the System with Elastic Elements of High Stiffness, *Izv. Akad. Nauk SSSR, Mekhan. tverd. tela*, No. 4, 1988, pp. 24–27.

47. Magnus K., *Kreisel Theorie und Anwendungen*, Springer-Verlag, Berlin–Heidelberg–New York, 1971 (in German).

48. Neymark Yu.I. and Fufaev N.A., *Dynamics of Nonholonomic Systems*, Nauka, Moscow, 1967 (in Russian).

49. Karapetyan A.V., About Realization of Nonholonomic Constraints and Stability of Celtic Stones, *Pricl. matem. i mekhan.*, Vol. 44, No. 1, 1981, pp. 42–51.

50. Kozlov V.V., Dynamics of System with Nonintegrable Constraints I, II, III, *Vestnic MGU*. Ser.1. Matem., mekhan., No. 3, 1982, pp. 92–100; No. 4, pp. 70–76; No. 3, 1983, pp. 102–111.

51. Kalinin V.V. and Novozhilov I.V., About Necessary and Sufficient Conditions of Nonholonomic Constraints Realizablility with Coulomb's Friction Forces, *Izv. Akad. Nauk SSSR, Mekhan. tverd. tela*, No. 3, 1975, pp. 15–19.

52. Novozhilov I.V., About Skidding while Braking, *Izv. Akad. Nauk SSSR, Mekhan.tverd. tela*, No. 4, 1975, pp. 15–19.

53. Lazaryan V.A., Dlugach L.R., and Karatenko M.L., *Rail-Vehicles Motion Stability*, Naukova dumka, Kiev, 1972 (in Russian).

54. Novozhilov I.V., Motions Decomposition of Rail-Vehicle, *Izv. Akad. Nauk SSSR, Mekhan. tverd. tela*, No. 4, 1983, pp. 66–70.

55. Novozhilov I.V., Control of Four-Axis Gimbal, *Izv. Akad. Nauk SSSR, Mekhan. tverd. tela*, No. 4, 1983, pp. 66–70.

56. Novozhilov I.V., Control of Legged-Vehicle Leg in Support Phase, *Biomekhanika*, Trudy Rizhsk. NII travm. ortoped, No. 13, 1975, pp. 634–639 (in Russian).

57. Bolotin Yu. V. and Novozhilov I.V., Control of Gait of Two-Legged Vehicle, *Izv. Akad. Nauk SSSR, Mekhan. tverd. tela*, No. 3, 1977, pp. 47–52.

58. Novozhilov I.V., Control of Space Motion of Two-Legged Vehicle, *Izv. Akad. Nauk SSSR, Mekhan. tverd. tela*, No. 4, 1984, pp. 47–53.

59. Merkin D.R., *Gyroscopic Systems*,Nauka, Moscow, 1974 (in Russian).

60. Novozhilov I.V., About Reducing Degree in Gyroscopic Systems of Equations, *Inzh. zhurn., Mekhan. tverd. tela*, No. 5, 1966, pp. 33–39.

61. Novozhilov I.V., About Application of Asymptotic Expansions of Theory of Differential Equations with Small Parameter by High Derivative for Gyroscopic Systems Research, *Izv. Akad. Nauk SSSR, Mekhan. tverd. tela*, No. 4, 1979, pp. 50–57.

62. Kobrin A.I., Martinesko Yu. G., and Novozhilov I.V., About Equations of Gyroscopic Systems, *Pricladn. Matem. i Mekhan.*, Vol. 40, No. 2, 1970, pp. 231–237.

63. Novozhilov I.V., About Angular Errors of Indicating Gyroscopic Stabilization System, *Izv. Akad. Nauk SSSR, Mekhan. tverd. tela*, No. 2, 1977, pp. 3–7.

64. Ishlinsky A.Yu., *Orientation, Gyroscopes and Inertial Navigation*, Nauka, Moscow, 1976.

65. Novozhilov I.V., About Stability of Three-Axis Gyrostabilizer, *Izv. Akad. Nauk SSSR, Mekhan.*, No. 5, 1965, pp. 137–140.

66. Novozhilov I.V., About Stability of Three-Axis Powered Gyroscopic Stabilizer, *Inzh. zhurn. Mekhan. tverd. tela*, No. 4, 1968, pp. 59–64.

67. Kobrin A.I. and Novozhilov I.V., About Stability of Two-Coordinate Automatic Following Tracing Systems, in: *Analiz i sintez sist. avtom. uprav.*, Moscow, 1968, pp. 382–384.

68. Novozhilov I.V., About possibility of separate investigation of "fast" and "slow" motion components in control problem of solid orientation with flywheels, *Kosmich. issled.*, Vol. 7, No. 5, 1977, pp. 693–700.

69. Ionkin I.L., Kus'mina R.P., and Novozhilov I.V., Stability of Multidimensional Control System of Dynamic Simulator, *Trudy Moskov. energ. in.*, No. 140, 1987, pp. 33–43.

70. Lewis A.E.D., Mathematical modelling of hemodiafiltration with particular reference to prediction of clearance. *Hemodiafiltration, Proceeding 1 Symposion Giessen, 1981*, Verlag Hygienerplan, Oberusel, pp. 63–75.

71. Novozhilov I.V., Kulakov G.P., and Balabanov P.I., About estimate of influence of blood and dializing-solution flows velocities and quality of semi-pervious membrane quantity on the efficiency of dialyzer, *Med. tekhn.*, No. 2, 1977, pp. 24–29.

72. Mazhdrakov G. and Popov N. (ed.), *Illnesses of Kidneys*, Meditsina i fizkult., Sofiya, 1969 (in Russian).

73. Ostoslavsky I.V., *Aerodynamics of Aircraft*, Gosud. izd. oboron. prom., Moscow, 1957 (in Russian).

74. Byushgens G.S. and Studnev P.V., *Aerodynamics of Aircraft. Dynamics of Longitudinal and Lateral Motion*, Mashinostroenie, Moscow, 1957 (in Russian).

75. Bochkarev A.F., et al., *Aeromechanics of Aircraft*, Mashinostroenie, Moscow, 1985 (in Russian).

76. Kuz'mak G.E., *Dynamics of Noncontrolled Flight Vehicles Motion by Entrance into Atmosphere*, Nauka, Moscow, 1970 (in Russian).

77. Borzov V.I., The Problem of Motion Decomposition in Flight Dynamics, *Izv. Akad. Nauk SSSR, Mekhan. tverd. tela*, No. 5, 1981, pp. 3–11.

78. Khachaturov A.A. (ed.), *Dynamics of System "Road-Tyre-Motor Car-Driver,"* Mashinostroenie, Moscow, 1978 (in Russian).

79. Ellis J.R., *Vehicle Dynamics*, Business Books Ltd., London, 1969.

80. Levin M.A., Fufaev N.A., *Theory of Deformable Wheel Rolling*, Nauka, Moscow, 1989 (in Russian).

81. Pacejka H.B., Lateral Dynamics of Road Vehicles. *Vehicle System Dynamics*, Vol. 16, 1987, pp. 75–120.

82. Ishlinsky A.Yu., About Sliding Within Contact Area with Rolling Friction, *Izv. Akad. Nauk SSSR, Otdel. tekhn. nauk*, No. 6, 1956, pp. 3–15 (in Russian).

83. Koz'ma Prutkov, *Composition*, GIKHL, Moscow, 1955.

Author Index

Subject Index